Einstein's Relativity in Great Britain

From Larmor and Eddington to Penrose
and Hawking: A Tale of Physicists, Astronomers,
Mathematicians and Philosophers

Einstein's Relativity in Great Britain

From Larmor and Eddington to Penrose
and Hawking: A Tale of Physicists, Astronomers,
Mathematicians and Philosophers

José M. Sánchez-Ron

Universidad Autónoma de Madrid, Spain
Real Academia Española, Spain

World Scientific

NEW JERSEY · LONDON · SINGAPORE · BEIJING · SHANGHAI · TAIPEI · CHENNAI

Published by

World Scientific Publishing Co. Pte. Ltd.
5 Toh Tuck Link, Singapore 596224
USA office: 27 Warren Street, Suite 401-402, Hackensack, NJ 07601
UK office: 57 Shelton Street, Covent Garden, London WC2H 9HE

Library of Congress Control Number: 2024950371

British Library Cataloguing-in-Publication Data
A catalogue record for this book is available from the British Library.

EINSTEIN'S RELATIVITY IN GREAT BRITAIN
From Larmor and Eddington to Penrose and Hawking: A Tale of Physicists, Astronomers,
Mathematicians and Philosophers

Copyright © 2025 by World Scientific Publishing Co. Pte. Ltd.

All rights reserved. This book, or parts thereof, may not be reproduced in any form or by any means, electronic or mechanical, including photocopying, recording or any information storage and retrieval system now known or to be invented, without written permission from the publisher.

For photocopying of material in this volume, please pay a copying fee through the Copyright Clearance Center, Inc., 222 Rosewood Drive, Danvers, MA 01923, USA. In this case permission to photocopy is not required from the publisher.

ISBN 978-981-12-0028-1 (hardcover)
ISBN 978-981-12-0029-8 (ebook for institutions)
ISBN 978-981-12-0030-4 (ebook for individuals)

For any available supplementary material, please visit
https://www.worldscientific.com/worldscibooks/10.1142/11278#t=suppl

Typeset by Stallion Press
Email: enquiries@stallionpress.com

Albert Einstein in a conversation with Arthur Eddington in the garden of the Cambridge Observatory. June 1930. Photograph attributed to Winifred Eddington. Reproduced in A. Vibert Douglas, *The Life of Arthur Stanley Eddington* (Thomas Nelson and Sons, 1956).

To Ana, Mireya, Amaya, Violeta and Tobías,
who give sense and purpose to my life.

Preface

One of the most interesting aspects of the history of science is the study of the origin and reception of theories which changed fundamental points and ideas of our scientific world view. Undoubtedly, one of those theories is Einstein's relativity, first (1905) the special theory, and after (1915) the general theory. In this book, I deal not with their origin but with their reception in Great Britain, where it finally found a sympathetic home. However, initially that reception was not easy, as the shadow of Maxwell was strong and many — among them Joseph Larmor, Lucasian professor at Cambridge University and one, though a rather peculiar, Maxwellian — thought that the basis for explaining experiments like the one carried out in 1887 by Michelson and Morley was Maxwell's electrodynamics and not a new theory, special relativity, which pretended to be more fundamental than it. Nevertheless, special relativity became accepted by a significant number of physicists, but it was general relativity which really opened the way to it in Great Britain, and elsewhere, more so after the announcement in November 1919 of the results of the British expedition to test part of the theory (the bending of light) during an eclipse of the Sun. Arthur Eddington, the leader of one the branches of that expedition, became the great defender of Einstein's relativity, an author of several books, especially his influential *The Mathematical Theory of Relativity* (1923).

Accordingly, the first two chapters of this book are dedicated to the early reception of relativity in Great Britain among physicists, as well as among mathematicians, who were interested due to the mathematical

x *Einstein's Relativity in Great Britain*

aspects of the theory. The third chapter deals with the reception among "the old guard" of physicists, those whose scientific world view was "classical physics", represented by Joseph Larmor and Oliver Lodge. Chapter four analyses how Einstein's relativity theories were received among British philosophers, among them Bertrand Russell and Alfred North Whitehead, the authors of the mythical, but finally failed, *Principia Mathematica* (1910–1913). The fifth chapter advances to a "new generation" of relativists, of whom I have selected George McVittie. And finally, chapter six has as its center a new way of understanding and working with relativity, the global structure of space-time, which had as its principal protagonists Roger Penrose and Stephen Hawking.

In the first five chapters I have used freely, though with many changes and additions, papers I have published in the past (Sánchez-Ron 1987a, 1987b, 1990, 1992, 1999, 2005, 2012).

Contents

Preface		ix
Chapter 1	The Reception of Special Relativity in Great Britain	3
Chapter 2	The Reception of General Relativity Among British Physicists and Mathematicians (1915–1930)	63
Chapter 3	Larmor, Lodge and General Relativity: The "Old Guard" Reaction to Einstein's Relativity	101
Chapter 4	The Early Reception of Relativity Among British Philosophers, or the Ductility of Philosophy	129
Chapter 5	A New Generation: George McVittie, "The Uncompromising Empiricist"	189
Chapter 6	The Renaissance of Relativity in Great Britain: From Coordinates to Global Space-Time and Black Holes	235
References		259

Joseph Larmor. Unknown author

Chapter 1

The Reception of Special Relativity in Great Britain

When Albert Einstein published in the *Annalen der Physik* his 1905 four epochal papers,[1] he was a completely unknown in the scientific community, just a technical expert at the Swiss Patent Office in Bern. He appeared as if out of the blue, something which adds interest to the question of how were received his revolutionary theories. In this book, I will deal only with the reception, among physicists, mathematicians, and philosophers, of his relativity theories, the special (1905) and the general (1915), in the case of Great Britain. That reception was more intense in Germany, where *Annalen der Physik* was published, and where one of its editors, Max Planck, supported Einstein's solution to the conflict between Newton's dynamics and Maxwell's electrodynamics. It was, indeed, more intense and also more diverse, with the participation of notable physicists as, among others, and besides Planck, Wilhelm Wien, Alfred Bucherer, Max von Laue, Jacob Laub, Arnold Sommerfeld, and the mathematician Hermann Minkowski, whose presentation of Einstein's theory unifying

---

[1]"Über einen die Erzeugung und Verwandlung des Lichtes betreffenden heuristichen Gesichtspunkt" (which enlarged Max Planck's 1900 introduction of quanta), "Über die von der molekularkinetischen Theorie der Wärme geforderte Bewegung von in ruhenden Flüssigkeiten suspendierten Teilchen" (where Einstein explained the Brwonian motion, and so supported molecular physics), "Zur Elektrodynamik bewegter Körper" (the special relativity paper), and "Ist die Trägheit eines Körpers von seinem Energieinhalt abhängig?" (the $E = mc^2$ paper).

4 *Einstein's Relativity in Great Britain*

space and time increased its revolutionary character.[2] On 3 October 1907, from Göttingen, where he was then professor of Mathematics, reclaimed by his friend David Hilbert, Minkowski, Einstein's former professor at the ETH in Zurich (he had followed eight courses with him), wrote to him (CPAE 1993: 77):

> "Dear Doctor Einstein:
>
> At our seminar [of partial differential equations of mathematical physics] in the W. S. we also wish to discuss your interesting papers on electrodynamics. If you still have available reprints of your article in the *Ann. D. Phys. U. Ch.*, Vol. 17 [actually, the name of the journal had been changed already in 1900 to *Annalen der Physik*], I would be grateful if you would send us a copy. I was in Zurich recently and was pleased to hear from different quarters about the great interest being shown in your scientific successes.
>
> With best regards, your sincerely
H. Minkowski."

And so, Minkowski developed his four-dimensional interpretation of special relativity, which he announced during his address at the 80th Assembly of German Natural Scientists and Physicians, at Cologne, on 21 September 1908, (Minkowski 1909: 104):

> "The views of space and time which I wish to lay before you have sprung from the soil of experimental physics, and therein lies their strength. They are radical. Henceforth space by itself, and time by itself, are doomed to fade away into mere shadows, and only a kind of union of the two will preserve an independent reality."

I

Before 1905: Maxwell's Shadow and the Maxwelians

1.1. The Maxwellians

In the case of Great Britain, the reception of the special theory of relativity was different, and, as we will see, slowest. One of the reasons was the

---

[2] For the reception of relativity in Germany, see Pyenson (1987).

deep influence which had what can be called "the Cambridge style in electrodynamics", a style dominated by James Clerk Maxwell's formulation of the equations of the electromagnetic field, one of the great achievements in science during the XIXth century, together with Darwin's theory of evolution, Gauss-Lobachevskii-Bolyai-Riemann's non-Euclidean geometry, thermodynamic, Pasteur and Koch's theory of germ disease, and the organic chemistry developments due to, among other, Berzelius, Wöhler and Liebig.

Maxwell's formulation reinforced the idea, introduced by Michael Faraday, that the electromagnetic force was transmitted through a continuous medium, an ether, or a *field*. In his address to the Mathematical and Physical Section of the 58th meeting of the British Association for the Advancement of Science held at Bath in September 1888, the Irish physicist George Francis FitzGerald (1851–1901), professor of Physics at Trinity College, Dublin, expressed well such conviction (FitzGerald 1889a: 558, 1902: 231):

> "In a presidential address on the borderlands of the known, delivered from this chair, the great Clerk Maxwell spoke of as an undecided question whether electromagnetic phenomena are due to a direct medium at a distance or are due to the action of an intervening medium. The year 1888 will be ever memorable as the year in which this great question has been experimentally decided by Hertz in Germany, and, I hope, by others in England. It has been decided in favour of the hypothesis that these actions take place by means of an intervening medium."

At the same meeting in Bath, Oliver Lodge, of whom much will be said later on, paid tribute to FitzGerald's role in establishing the ether on solid grounds (quoted in Lodge 1931: 101–102):

> First, then, the section was happy for its president [FitzGerald] a man exhibiting an altogether exceptional combination of qualities — profound knowledge, great originality, mathematical and experimental skill both of the first order, quickness of comprehension, and a ready wit. The address with which opened the business of the section was not unworthy of the man. Its subject was that greatest of all physical subjects, 'The Ether,' and it largely consisted in a glorification of the quite recent experiments of Dr. Hertz en Berlin, whereby Prof. FitzGerald shows that

6 Einstein's Relativity in Great Britain

the existence of an ether is raised out of the rank of a very probable hypothesis, which it has long been, into the dominion of demonstrated facts. He hints also that certain outstanding problems concerning the structure and properties of the ether can most probably be answered by a further discussion and repetition of experiments. Our engineering friends ought not to sneer at the prominence given to such refined abstraction or hypothetical figment as the ether may seem to some of them to be, for it is by its means they drive all electrical motors, it is it they use in dynamos and transformers of all sorts; it is the ether which transmits their signals, whether telegraphic or telephonic; and, though this may appeal to them less forcibly, it is through the action of the ether that mankind is enabled to see. Not to mention the as yet incompletely verified hypothesis that it is of ether that Engineers and all the other material substances are composed."

As we shall see later, Lodge's manifestation represented well the general feeling in Britain as regards the ether-field concept, not to mention Lodge's own ideas, which we will meet also later.

However, and though Maxwell unified electricity, magnetism and optics in four vector partial differential equations, it left many questions unsolved, which a group of British physicists, among them FitzGerald, Oliver Heaviside and Joseph Larmor, tried to solve during the 1880s and 1890s, clarifying and extending the fundamental but difficult Maxwell's 1873 *Treatise on Electricity and Magnetism*. They were called, "the Maxwellians" (Hunt 1991).

Prominent among such questions was how the electromagnetic field, a medium — or ether — viewed as distinct from matter, interacts with the small charged particles (electrons), which many assumed to form part of the entities, the atoms, which made up matter.[3] And how to solve that question in such a way that the negative result of Michelson and Morley's 1887 experiments were explained (Michelson and Morley 1887). The problem was that there were no effects in the motion of the Earth through the ether, effects which were measurable using the velocity of light.

---

[3] It was only in 1897 that J.J. Thomson discovered the electron as a basic part of matter.

1.2. Joseph Larmor, a Different Maxwellian

Another prominent Maxwellian, though of a different class, was the distinguished mathematical physicist Joseph Larmor (1857–1942), to whom chapter 3 will be dedicated. At Cambridge University, he took the Tripos examinations in 1880, and came out as Senior Wrangler, with J.J. Thomson being Second Wrangler. After becoming a lecturer at St. John's College, Cambridge, in 1903 he was chosen to succeed Gabriel Stokes as Lucasian professor, a chair which he retained until 1932. Knighthood in 1909, he was president of the London Mathematical Society in 1914–1915, the Royal Society awarded him its Royal Medal in 1915 and in 1921 the Copley Medal; he also represented Cambridge University in Parliament from 1911 to 1922, being a firm conservative, in politics as in science.

Larmor's scientific work centred on the electromagnetic theory, optics, analytical mechanics, and geodynamics.[4] Among his most famous works figures his book, *Aether and Matter* (Larmor 1900a), significantly subtitled: "*A development of the dynamical relations of the aether to material systems on the basis of the atomic constitution of matter, including a discussion of the influence of the earth's motion on optical phenomena.*" According to Eddington (1942–1944: 198), this work "ranks among the great scientific books." It was certainly, the last of his creative and influential works.

Aether and Matter was a book which "helped to establish a research school that guided the development of mathematical electromagnetic theory in Cambridge until the end of War World I" (Warwick 1993: 49), a research school which "marked an important break, both conceptually and geographically, with the British Maxwellian tradition of the 1880s and early 1890s. From a theoretical perspective, the ETM [Electronic Theory of Matter] transcended the work of Thomson, Poynting, and others by positing the electron as the natural origin and unit of positive and negative electric charge and by offering a new unification of optics, electrodynamics, and matter theory. The ETM embodied what shortly became known as the 'Lorentz transformations' and relied heavily upon the null result of

---

[4]Larmor's papers are included, with a few exceptions (among which are, as I will explain in chapter 3, his papers on the theory of relativity) in Larmor (1929a, 1929b).

8 *Einstein's Relativity in Great Britain*

Michelson-Morley experiment to provide empirical support for its central claims that matter was electrically constituted, that all matter was electromagnetic in origin, and that matter contracted minutely in its direction of motion through the ether" (Warwick, 2003: 375). However, Larmor believed that the transformations he obtained were only empirically justified to the second order, the ones he included in *Aether and Matter*. It was only after Lorentz introduced the exact transformations in 1904 that he accepted their exact value, but the position he took in *Aether and Matter*, and the complication of his approach made that though recognised as a great contribution to electrodynamics, the influence of the book was not great. As to the influence of Lorentz in Larmor's work, what can briefly be said is what he wrote in *Aether and Matter* (Larmor 1900a: 18–19):

> "[The] necessary amendment of the scheme of Maxwell has been independently arrived at by more than one writer, but somewhat earliest in point of time by H.A. Lorentz; and it involves the general electrodynamic considerations, including the discrete distribution of electricity among the molecules of matter, on which the present essay is based."

An important point is that, as the title of his book suggested, Larmor thought that problems like the explanation of Michelson's result reinforced the idea that an ether was necessitated. Thus, in *Aether and Matter*, he wrote (Larmor (1900a: 186–187):

> "In favour of the view that the interactions between atoms are in very great part those necessitated by the aether whose properties are revealed in electric and optical phenomena, there is, in addition to the inherent theoretical difficulty in conceiving any other kind of interaction, the actual fact that on the lines of the above argument such a view does account for a definite and well-ascertained experimental result, that of Michelson, above discussed, which has hitherto stood by itself as the only quantitative observational evidence that has a bearing on this question."

However, such conclusion only affected the phenomena covered by Maxwell's electrodynamics, gravitation was another case (Larmor 1900: 187):

"It can be said on the other side that this view of aethereal action does not directly cover gravitational phenomena, unless the rather artificial pulsatory theory of gravity is allowed."

From what I have said so far, it can be deduced that Larmor was just a physicist of the old school (more about this in Chapter 3), a physicist who only felt at home with 19th century physics, a conservative individual. Thus, in the obituary of Larmor which he wrote for the Royal Society, Arthur Eddington (1942–44: 205) said that he "seemed a man whose heart was in the nineteenth century, with the names of Faraday, Maxwell, Kelvin, Hamilton, Stokes ever on his lips." Nevertheless, Larmor's physics depended much on a rather abstract element, the Least Action Principle, so much that he believed in its fundamental character that had he been faced with the pressing necessity of choosing between the ether and the Least Action Principle, he would have opted for the latter; that is, he would have behaved in what can be considered a "modern" manner.[5] Take, for instance, the following passage from his Presidential Address, "The methods of mathematical physics," to the Mathematical and Physical Science Section of the British Association for the Advancement of Science meeting held in Bradford during September 1900 (Larmor 1900b: 620–621; 1929b: 204):

"But whatever views may be held as to the ultimate significance of this principle of Action, its importance not only for mathematical analysis, but as a guide to physical exploration, remains fundamental. When the principles of the dynamics of material systems are refined down to their ultimate common basis, this principle of minimum is what remains. [...] In so far as we are given the algebraic formula for the time-integral which constitutes the Action, expressed in terms of any suitable coordinates, we know implicitly the whole dynamical constitution and history of the system to which it applies. Two systems in which the Action is expressed by the same formula are mathematically identical, are

---

[5]According to Eddington (1941–1944: 204): "Larmor had an intense, almost mystical, devotion to the principle of least action; [to him] it was the ultimate natural principle — the mainspring of the universe." About Larmor's arguments in favour of the fundamental character of the Least Action Principle, see also Larmor (1884).

physically precisely correlated, so that they have all dynamical properties in common. When the structure of a dynamical system is largely concealed from view, the safest and most direct way towards answering the prior question as to whether it is a purely dynamical system at all, is thorough this order of ideas. *The ultimate test that a system is a dynamical one is not that we shall be able to trace mechanical stresses throughout it, but that its relations can be in some way or other consolidated into accordance with this principle of minimum Action* [emphasis added].

However, it is true that ether and the Least Action Principle seemed to get along quite well. In Larmor's (1900b: 622; 1929b: 205–206) own words:

"Returning to the molecules, it is now verified that the Action Principle forms a valid foundation throughout electrodynamics and optics; the introduction of the aether into the system has not affected its application. It is therefore a reasonable hypothesis that the principle forms an allowable foundation for the dynamical analysis of the radiant vibrations in the system formed by a single molecule and surrounding ether."

I will say more about Larmor and the Least Action Principle in Chapter 3.

1.3. H. A. Lorentz and G. F. FitzGerald

The Dutch physicist Hendrik Antoon Lorentz (1853–1928) was the more prominent in the efforts to explain such "anomaly." Starting with his doctoral thesis, *Over de theorie der terugkaatsing en breeking van het licht* (*On the theory of reflection and refraction of light*; 1875), Lorentz dedicated himself to develop Maxwell's electrodynamics. We must take into account that at first it was not clear that Maxwell's theory would be correct, as Lorentz (1875, 1935: 75) recognized in his thesis when he said: "The equations from the previous chapter will serve us to study the remarkable discovery by which Maxwell was led to consider light as an electromagnetic phenomenon and to develop a theory, which, if it proves to be correct, will be one of the most beautiful examples of the happy results that can provide the mathematical study of natural phenomena."

In his long march towards solving the above-mentioned problem of the motion of the Earth through the ether, Lorentz a few articles stand out. One of them, "The relative motion of the Earth and the Ether," was published in 1892. I will quote from it (Lorentz 1892, 1937a: 219–221):

"In order to explain the aberration of light, Fresnel [which Lorentz admired] assumed that the ether does not partake of the yearly motion of the Earth, which, naturally also means that our planet is perfectly permeable to this medium. Later on Stokes attempted another explanation by supposing the ether to be dragged along by the Earth and that, consequently, at every point of the Earth's surface the velocity of the ether is equal to that of the Earth. [...]

A serious difficulty however had arisen in an interference experiment made by Michelson [*Amer. Journal of Science 22*, 120 (1875)] in order to make a decision between the two theories.

Maxwell had already observed that if the ether is not dragged along, the motion of the Earth must influence the time required by light to travel to and fro between two points rigidly fixed to the Earth [...]. Michelson made use of an apparatus with two horizontal arms of an equal length and perpendicular to each other, supporting at their ends mirrors at right angles to their directions. An interference phenomenon was observed while the one beam of light was traveling from the point of intersection of the arms to and fro along the one arm, and the second beam along the other. The whole apparatus, including the source of light and the observing telescope, could be rotated on a vertical axis; also, the phenomenon was observed at such a time as to permit the best possible adjustment of either of the arms in the direction of the Earth's motion. [...] Not the slightest shift, however, of the interference-fringes could be detected. [...]

This experiment has been puzzling me for a long time, and in the end I have been able to think of only one means of reconciling its results to Fresnel's theory. It consists in the supposition that the line joining two points of a solid body, if at first parallel to the direction of the Earth's motion, does not keep the same length when it is subsequently turned through 90°.

Now, some such change in the length of the arms in Michelson's first experiment and in the dimensions of the slab in the second one is so far as I can see, not inconceivable. What determines the size and shape

12 Einstein's Relativity in Great Britain

of a solid body? Evidently the intensity of the molecular forces; any cause which would alter the latter would influence the shape and dimensions. Nowadays we may safely assume that electric and magnetic forces act by means of the intervention of the ether. It is not far-fetched to suppose the same to be true of the molecular forces. But then it may make all the difference whether the line joining two material particles shifting together thorough the ether, lies parallel or crosswise to the direction of that shift."

As a matter of fact, the same idea had occurred to FitzGerald, who presented it in 1889 in a *Letter to the Editor*, sent on May 2, from Dublin, "The Ether and the Earth's Atmosphere," published in *Science* (Vol. 13, 390), the journal of the American Association for the Advancement of Science, founded in 1848.[6] As it is short, and it was not included in *The Scientific Writings of the Late George Francis FitzGerald* (1902), edited, it is interesting to note, by Joseph Larmor, I will reproduced it in its entirety (FitzGerald 1889b: 390):

"I have read with much interest Messrs. Michelson and Morley's wonderfully delicate experiment attempting to decide the important question as to how far the ether is carried along by the Earth. Their result seems opposed to other experiments showing that the ether in the air can be carried along only to an inappreciable extent. I would suggest that almost the only hypothesis that can reconcile this opposition is that the length of material bodies changes, according as they are moving through the ether or across it, by an amount depending on the square of the ratio of their velocity to that of light. We know that electric forces are affected by the motion of the electrified bodies relative to the ether, and it seems a not improbable supposition that the molecular forces are affected by the motion, and that the size of a body alters consequently. It would be very important if secular experiments on electrical attractions between permanently electrified bodies, such as in a very delicate quadrant electrometer, were instituted in some of the equatorial parts of the Earth to observe whether there is any diurnal and annual variation of attraction — diurnal due to the rotation of the Earth being added and

---

[6] On the origin of FitzGerald's idea see Hunt (1988).

subtracted from its orbital velocity; and annual similarly for its orbital velocity and the motion of the solar system."

FitzGerald and Lorentz's ideas about the length contraction were made independently. Indeed, when Lorentz learnt that FitzGerald had advanced the same idea, he wrote him the following letter on November 10, 1894 (Kox 2008: 45):

"My dear Sir,

Prof. Oliver Lodge, in his 'Aberration Problems' [Lodge (1893)] mentions a hypothesis, which you have imagined in order to account for the negative result of Mr. Michelson's experiment. Some time ago I arrived at the same view, as you may see from the number of the Proceedings of the Dutch Academy of Sciences which I have the honour to forward to you conjointly with this letter [Lorentz (1892)].

A memoir in which I consider the whole subject of Aberration in connection with the electromagnetic theory of light being now in course of publication [Lorentz (1895)] — it will in fact appear in a week —, you would oblige me very much by telling me, if your hypothesis has been published. I have been unable to find it and yet I should wish to refer to it.

Most respectfully yours,
H. A. Lorentz."

Four days after, that on November 14, FitzGerald answered (Kox 2008: 46):

"My dear Sir,

I have been for years preaching and lecturing on the doctrine that Michelson's experiment proves, and is one of the only ways of proving, that the length of a body depends on how it is moving through the ether. A couple of years after Michelson's results were published, as well as I recollect, I wrote a letter to 'Science', the American paper that has recently become defunct, explaining my view, but I do not know whether they ever published it, for I did not see the journal for some time afterwards.[7] I am pretty sure that your publication is then prior to any of my

---

[7]This explain why the paper did not appear in *The Scientific Writings of the Late George Francis FitzGerald* (1902).

14   *Einstein's Relativity in Great Britain*

printed publications for I have looked up several places where I thought I might have mentioned it but cannot find that I did. I certainly ever wrote any special article about it as I ought to have done for the information of others besides my students here.

I am particularly delighted to hear that you agree with me, for I have been rather laughed at my view over here. I could not even persuade my own pupil W. Preston to introduce this criticism into his book on Light published in 1890 [Preston (1890)] although I pressed him to do so and it was only after reiterated positiveness that I induced Dr. Lodge to mention it in his paper [Lodge (1893)]; but now that I have you an advocate and authority I shall begin to jeer at others for holding any other view.

Thank you very much for your papers. I can make out their general drift and wish I were able to reciprocate by replying to you in Dutch.

Yours most sincerely
Geo. Fras. Fitzgerald."

In a paper Lorentz published in 1897, "Concerning the problem of the dragging along of the ether by the Earth", he recognized FitzGerald's contribution (Lorentz 1897, 1937a: 241):

"One or two difficulties remain. One lies in an earlier experiment of Michelson and Morley, in which two rays of light interfered which, one in one direction, the other in a direction perpendicular thereto, moved forwards and backwards over a given distance. It appeared that the motion of the Earth had no influence on the position of the interference bands observed under these conditions.

To explain this result I have made the following hypothesis which also occurred to FitzGerald."

And here he explained the length contraction hypothesis.

Lorentz called his efforts to solve the problems that affected Maxwell's electrodynamics the "Theory of the electron", an "extension to the domain of electricity [in Maxwell's sense] of the molecular and atomistic theories that have proved of so much use in many branches of physics and chemistry," as he wrote in his book *The Theory of Electrons* (Lorentz 1909, 1952: 10), the result of the lectures he delivered in Columbia University (New York) in the spring of 1906.

It was in a paper published in 1904 that Lorentz arrived to what will called in his honour, when included in Einstein's 1905, "the Lorentz's transformations" of space and time. The title of the paper was "Electromagnetic phenomena in a system moving with any velocity smaller than that of light". There, Lorentz (1904, 1937b: 172) paid tribute to FitzGerald's idea:

> "The problem of determining the influence exerted on electric and optical phenomena by a translation such as all systems have in virtue of the Earth's annual motion, admits of a comparatively simple solution, so long as only those terms need be taken into account, which are proportional to the first power of the ratio between the velocity of translation w and the velocity of light c. Cases in which quantities of the second order, i.e., of the order w^2/c^2, may be perceptible, present more difficulties. The first example of this kind is Michelson's well known interference-experiment, the negative result of which has led Fitz Gerald and myself to the conclusion that the dimensions of solid bodies are slightly altered by their motion through the aether."

II

The Ether, a Concept Appreciated in Britain, but not by All

1.4. The Ether

Before arriving to the effective reception of special relativity in Great Britain, it is necessary to consider the status that the ether concept had there, as one of the points Einstein made in his revolutionary 1905 article was that "the introduction of a 'luminiferous ether' will prove to be superfluous in as much as the view here to be developed will not require an 'absolute stationary space' provided with special properties, nor assign a velocity-vector to a point of the empty space in which electromagnetic processes take place."

Long ago, Stanley Goldberg (1970, 1984) argued that there was widespread acceptance of the ether concept among British physicists during the nineteenth century and first decades of the twentieth, something, of

16 *Einstein's Relativity in Great Britain*

course, not surprising if we take into account the influence of the Maxwellians. One of the consequences of that situation was, wrote Goldberg (1984: 221), "that the acceptance [of special relativity] hinged upon making it compatible with the concept of the ether. As paradoxical as that might be, there was almost unanimous agreement within the British physicist community about such a program."

There are many examples in this sense. J.J. Thomson (1856–1940), the director of the Cavendish Laboratory (Cambridge) and discoverer of the electron as a subatomic particle, provides one of them. One particularly interesting is contained in the presidential address he delivered during the Seventy-ninth meeting of the British Association for the Advancement of Science which took place in Winnipeg, August 25–September 1, 1909. There Thomson (1910: 15) said:

"The matter of which I have been speaking so far is the material which builds up the earth, the sun and the stars, the matter studied by the chemist, and which he can represent by a formula; this matter occupies, however, but an insignificant fraction of the universe, it forms but minute islands in the great ocean of the ether, the substance with which the whole universe is filled.

The ether is not a fantastic creation of the speculative philosopher; it is as essential to us as the air we breathe. For we must remember that we on this earth are not leaving on our own resources; we are dependent from minute to minute upon what we are getting from the sun, and the gifts of the sun are conveyed to us by the ether. It is to the sun that we owe not merely night and day, springtime and harvest, but it is the energy of the sun, stored up in coal, in waterfalls, in food, that practically does all the work of the world."

In his autobiography, *Recollections and Reflections*, J.J. Thomson (1936: 431–433) related his ideas with Einstein's new theory of relativity:

"The theory of relativity deals with physical phenomena. If we take the view that the structure of matter is electric, these ought to follow from Maxwell's equations without introducing relativity. We have referred to examples in which various effects have been explained in this way, e.g. the contraction of a moving body and the variation of mass with velocity, before relativity was introduced. On this view it is reasonable to regard Maxwell's equations as the fundamental principle rather than that of relativity, and also to regard the ether as the seat of the mass momentum and energy of matter, i.e. of protons and electrons: lines of force being the bounds which bind ether to matter. In Einstein's theory there is no mention of an ether, but a great deal about space: now space if it to be of any use in physics must have much the same properties as we ascribe to the ether; for example, as Descartes pointed out long ago, space cannot be a void. Unless there was something in space there would be nothing to fix the position of a point, position would have no meaning and space could have no geometrical properties; the same would be true even if it were filled with a perfectly uniform substance, for again there would be nothing to differentiate on point from another. Space must therefore have a structure. Again, there must be in space something which changes. If there were not there would be nothing to distinguish one instant from another, nothing to supply a 'clock'. Nothing can travel through space with a velocity greater than light; this is the speed limit for traffic through it, but space could not enforce this unless it had some time to measure the velocity, and this requires something by which we could measure times, Again, the mass of a body increases as the velocity increases; if the mass does not come from space it must be created. It would seem that space must possess mass and structure both in time and space: In fact it must possess the qualities postulated for the ether:

'Naturam expellas furca, tamen usque recurret'

It has been urged against the existence of the ether that so many different kinds of ether have been from time to time proposed. The same may be said about 'space': We have Einstein's space, de Sitter's space, expanding universes, contracting universes, vibrating universes, mysterious universes. In fact the pure mathematicians may create universes just by writing down an equation, and indeed if he is an individualist he can have a universe of his own.[8]

---

[8] These examples belong to the general theory of relativity, not to the special theory. About the different classes of ether, see Schaffner (1972) and Cantor and Hodge, eds. (1981).

The kind of relativity I have considered, 'special relativity' as it is now called, was Einstein's first theory and deals with electric and magnetic problems which can also be solved by Maxwell's equations. Einstein has given a second theory, known as 'general relativity', which includes a theory of gravitation. This involves much very abstruse and difficult mathematics, and there is much of it I do not profess to understand. I have, however, a profound admiration for the masterly way in which he has attacked a problem of transcendent difficulty."

J.J. Thomson's case is but one example of a distinguished physicist who was slow in fully accepting the consequences of Einstein's 1905 theory of relativity. Were I to pay more attention to those British scientists who were reluctant to accept Einstein's theory, I would mention, among others, the great Ernest Rutherford. His friend, the British physicist, educated in Cambridge, but who carried out most of his career at McGill University (Montreal, Canada), Arthur Stewart Eve (1939: 193) remembered that during a conversation held in 1910 with Wilhelm Wien and Rutherford, Wien manifested that "no Anglo-Saxon can understand relativity," to which Rutherford replied "No!, they have too much sense." And, independently of its possible jocular character, Rutherford's answer might be considered as representative of a large group of British physicists. The motivations and degrees of understanding of Einstein's restricted theory were, of course, different among all these opponents to special relativity, but their very existence and importance can hardly be denied.

Take, for example, the case of Arthur Schuster, professor and director of the Physics Laboratory at Manchester University (strictly them, Victoria University) until 1907, when he retired and Rutherford took his post. By 1908, Schuster understood well the meaning and consequences of special relativity, but, however, had doubts about its correctness, as he stated in his book *The Progress of Physics during 33 years (1875–1908)* (Schuster 1911: 108–111):

"the conclusion finally arrived at by Michelson and Morley was, that so far as their experiments could decide, the difference in path, calculated on the hypothesis of a quiescent aether did not exist; but a quiescent aether is one of the few necessities of modern physics. How are we to meet this difficulty? Mathematicians are always sufficiently resourceful to cope with any problem set to them by the experimentalist, and as Poincaré likes to tell us, we can always find a new hypothesis to fit a new fact. The hypothesis in the present instance was supplied independently by FitzGerald and Lorentz. [...] Further, Lorentz and Larmor have given grounds for believing that if molecular forces have ultimately an electric origin, such a contraction ought to take place. The matter did not rest there, mainly perhaps because the explanation was felt by some to be a little artificial. [...] Einstein, in a paper of great interest and power, has developed this idea, calling his imagined law 'The principle of relativity,' because it stipulates — *a priori* — that only relative motion between material bodies can be detected. It is impossible for me to discuss in detail the reasoning by which this principle is justified, and an account by this without explanations of its consequences would lay open to the charge that I was playing with your credulity. Suffice, therefore, it to say that strict adherers to the principle cannot admit the existence of an aether, and yet may speak of the transmission f light through space with a definite velocity. [...] The theory appears to have an extraordinary power of fascinating mathematicians, and it will certainly take its place in any critical examination of our scientific beliefs; but we must not let the simplicity of the assumption underlying the principle hide the very slender experimental basis on which it rests at present, and more especially not lose sight of the fact, that it goes much beyond what is proved by Michelson's experiment. [...] Einstein's generalisation assumes that the result of the experiment would still be the same, if performed in free space with the source of light and mirrors disconnected from each other but endowed with a common velocity. This is a considerable and, perhaps, not quite justifiable generalisation. I am well aware that Bucherer's experiments with kathode rays are taken to confirm the validity of Einstein's principle, but if we say that they are not inconsistent with it, we should probably go as far as s justifiable."

Even as late as 1927, it can be hear voices reminiscences of the old ether. So, it was with Edmund Whittaker (1927: 16–17), of whom more will be said in other chapters. Speaking about "The outstanding problems of Relativity" as president of the Section A ("Mathematics and Physical Sciences") during the 95 meeting of the British Association for the Advancement of Science (Leeds, August 31–September 7)

20 *Einstein's Relativity in Great Britain*

"It was January 1914 that Einstein made his great departure from the Newtonian doctrine of gravitation by abandoning the idea that the gravitational potential is scalar. The thirteen eventful years which have passed since then have seen the rapid development of the new theory, which is called General Relativity, and the confirmation by astronomers and astrophysicists of its predictions regarding the bending of light by the sun and the displacement of spectral lines. At the same time s number of new problems have arisen in connection with it; and perhaps the time has now come to review the whole situation and to indicate where there is need for further investigation.

Speaking from this Chair I may perhaps be permitted to recall that my first experience of the British Association was as one of the secretaries of Section A nearly thirty years ago; and that my secretarial duties brought me the privilege of an introduction to the distinguished mathematical physicist, Prof. G. F. FitzGerald of Dublin, who was a regular and prominent member of the section until his death in 1901. FitzGerald had long held an opinion which he expressed in 1894 in the words 'Gravity is probably due to a change of structure if the aether, produced by the presence of matter.' Perhaps this is the best description of Einstein's theory that can be given in a single sentence in the language if the older physics: at any rate it indicates the three salient principles, firstly, that gravity is not a force acting as a distance, but an effect due to the modification of space (or, as FitzGerald would say, of the aether) in the immediate neighbourhood of the body acted on; secondly, that the modification is propagated from point to point of the space, being ultimately connected in a definite way with the presence of material bodies; and thirdly, that the modification is not necessarily of a scalar character. The mention of the aether would be criticised by many people to-day as something out of date and explicable only by the circumstance that FitzGerald was writing thirty years ago; but even this criticism will be not universal; for Wiechert and his followers have actually combined the old aether theory with ideas resembling Einstein's by the hypothesis that gravitational potential is an expression of what we may call the specific inductive capacity and permeability of the aether, those qualities being affected by the presence of gravitating bodies. Assuming that matter is electrical in its nature, it is inferred that will be attracted to places of greater dielectric constant. It seems possible that something of this sort was what FitzGerald had in mind."

1.5. Oliver Lodge, and "the Ether of Space"

Were I to choose a radical, and moving, statement in favour of the ether concept, I will select the "Foreword" of a book, *My Philosophy*, significantly subtitled *Representing My Views on the Many Functions of the Ether of Space*, who Oliver Lodge (1933: 5), a prominent British physicist of whom I will say more in Chapter 2:

> "The Ether of Space has been my life study, and I have constantly urged its claims to attention. I have lived through the time of Lord Kelvin with his mechanical models of the ether, down to a day when the universe by some physicists seems resolved into mathematics, and the idea of an ether is by them considered superfluous, if not contemptible. I always meant some day to write a scientific treatise about the Ether of Space; but when in my old age I came to write this book, I found that the Ether pervaded all my ideas, both of this world and the next. I could no longer keep my treatise within the proposed scientific confines; it escaped in every direction, and now I find has grown into a comprehensive statement of my philosophy."

As a matter of fact, Lodge's statement is but a manifestation of his belief in a world which laid beyond the observable reality. His ether fitted well with what Denis Weaire (2009: 133) wrote: "The mysterious ether was eagerly adopted by the spiritualists who became fashionable in the Victorian period, as a pseudoscientific justification of their claims."

1.6. Joseph Henry Poynting's Ideas About the Ether

Though much of what Goldberg said is true, and independently of what physicists like J.J. Thomson, whose work was not as related to electrodynamics and Larmor or other Maxwellians, there existed also other trends within British physics that, when taken into account, present a more diversified picture that the one which is dominated by the ether concept, and John Henry Poynting (1852–1914) was a good example in this sense.

Poynting spent the largest part of his career in Birmingham, first as professor of physics at Mason College, and later on, when in 1900 Mason

22   *Einstein's Relativity in Great Britain*

College became the University of Birmingham, as dean of the Faculty of Science till 1912. Among his writings, there is one particularly germane to my purposes: his Presidential Address to the Mathematical and Physical Section of the British Association for the Advancement of Science (BAAS) meeting held in Dover in 1899. As president of that section, he had to survey its field, and because of this he paid attention to concepts like the ether (Poynting 1920: 605):

> "While light was regarded as corpuscular — in fact molecular — and while direct action at a distance presented no difficulty, the molecular hypothesis served as the one foundation for the mechanical representation of phenomena. But when it was shown that infinitely the best account of the phenomena of light could be given on the supposition that it consisted of waves, something was needed, as Lord Salisbury has said, to wave, both in the interstellar and in the intermolecular spaces. So the hypothesis of an ether was developed, a necessary complement of that form of the molecular hypothesis in which matter consists of discrete particles with matter-free intervening spaces."
>
> At this point, the questions which Poynting (1920: 606) faced were:
> "How are we to regard these hypotheses as to the constitution of matter and the connecting ether? How are we to look upon the explanations they afford? Are we to put atoms and ether on an equal footing with the phenomena observed by our senses, as truths to be investigated for their own sake? Or are they mere tools in the search for truth, liable to be worn out or superseded?"

To answer such questions, he began by directing his attention to rather elemental facts:

> "That matter is grained in structure is hardly more than the expression of the fact that in very thin layers it ceases to behave as in thicker layers. But when we pass on from this general statement and give definite form to the granules or assume definite qualities to the intergranular cement we are dealing with pure hypothesis.
>
> It is hardly possible to think that we shall ever see an atom or handle the ether. We make no attempt whatever to render them evident to the senses. We connect observed conditions and changes in gross visible matter by invisible molecular and ethereal machinery. The changes at

each end of the machinery of which we seek to give an account are in gross matter, and this gross matter is our only instrument of detection, and we never receive direct sense impressions of the imagined atoms or the intervening ether. To a strictly descriptive physicist their only use and interest would lie in their service in prediction of the changes which are to take place in gross matter."

His conclusions were rather definite and, in some senses, probably shocking to many among his audience:

"It appears quite possible that various types of machinery might be devised to produce the known effects. The type we have adopted is undergoing constant minor changes, as new discoveries suggest new arrangements of the parts. Is it utterly beyond possibility that the type itself should change?

The special molecular and ethereal machinery which we have designed, and which we now generally use, has been designed because our most highly developed sense is our sense of sight. Were we otherwise, had we a sense more delicate than sight, one affording us material for more definite mental presentation, we might quite possibly have constructed very different hypotheses."

In spite of all these critical statements, Poynting (1920: 607) knew too well that: "It is merely a true description of ourselves to say that we must believe in the continuity of physical processes, and that we must attempt to form mental pictures of those processes the details of which elude our observation. For such pictures we must frame hypotheses, and we have to use the best material at command in framing then. At present there is only one fundamental hypothesis — the molecular and ethereal hypothesis — in some such form as is generally accepted." Nevertheless, even accepting this, he wished to point out what was, according to him, the real position of the ether concept, a position which was far from being the same as the one accepted by men like Lord Kelvin, Oliver Lodge, or George Green, to name but a few. And he added (Poynting 1920: 607–608):

"In this country there is no need for any defence of the use of the molecular hypothesis. But abroad the movement from the position in

24 *Einstein's Relativity in Great Britain*

which hypothesis is confounded with observed truth has carried many through the position of equilibrium equally far on the other side, and a party has been formed which totally abstains from molecules as a protest against immoderate indulgence in their use. [...]

But the protest will have value if it will put us on our guard against using molecules and the ether everywhere and every when. There is, I think, some danger that we may get so accustomed to picturing everything in terms of these hypotheses that we may come to suppose that we have no firm basis for the facts of observation until we have given a molecular account of them, that a molecular basis is a firmer foundation than direct experience. [...]

There is more danger of confusion of hypothesis with fact in the use of the ether: more risk of failure to see what is accomplished by its aid. In giving an account of light, for instance, the right course, it appears to me, is to describe the phenomena and lay down the laws under which they are grouped, leaving it an open question what it is that waves, until the phenomena oblige us to introduce something more than matter, until we see what properties we must assign to the ether; properties not possessed by matter, in order that it may be competent to afford the explanations we seek. We should then realise more clearly that it is the constitution of matter which we have imagined, the hypothesis of discrete particles, which obliges us to assume an intervening medium to carry on the disturbance from particle to particle."

There is another argument used by Poynting, which shows that he was "modern enough" as to be able to accept that there is no necessity of an ether for fixing the positions of material bodies, a question that, as we know, is closely related to the foundations of special relativity. The argument goes as follows in his own words (Poynting 1920: 610):

"Another illustration of the illegitimate use of our hypothesis, as it appears to me, is in the attempt to find in the ether a fixed datum for the measurement of material velocities and accelerations, a something in which we can draw our coordinate axes so that they will never !urn a bend. But this is as if, discontented with the movement of the earth's pole, we should seek to find our zero lines of latitude and longitude in the Atlantic Ocean. Leaving out of sight the possibility of ethereal currents which we cannot detect, and the motions due to every ray of light

The Reception of Special Relativity in Great Britain 25

which traverses space, we could only fix positions and directions in the ether by buoying them with matter. We know nothing of the ether, except by its effects on matter, and, after all it would be the material buoys which would fix the positions and not the ether in which they float."

1.7. Against the Ether: Norman R. Campbell (1)

Norman Robert Campbell (1880–1949) was a physicist who studied at Trinity College, Cambridge, and got his B.A. in 1902, after which he became fellow at Trinity College and research assistant at the Cavendish Laboratory under the direction of J.J. Thomson. Afterwards, he carried out his career at Leeds University as honorary fellow in physics research at the National Physics Laboratory, and the Hirst Research Centre of the General Electric Company; besides, he had interests in the philosophy of science, to which he contributed with a book: *What is Science* (1921). Though his career as scientist was not too prominent, he wrote a series of influential books, of which the most well-known work was his book *Modern Electrical Theory*, which went through two editions (Cambridge University Press, 1907, 1913), the second one enlarged with several appendixes, almost monographs if taken by themselves.[9]

In the first edition, in which there were no references to Einstein's 1905 article, Campbell avoided almost completely any reference to the ether concept until the very last chapter. There he began by referring (Campbell 1907: 288) to the many "confusions and misunderstandings" which affect it. "The amazing pronouncements about the 'ether' which have been made by many philosophers are rivalled by the statements which are to be found in the writings of men of science of the highest repute." As a matter of fact, Campbell was not "acquainted with any authoritative formal definition of the term 'aether'," and consequently he took the course, "not diverging from current usage," of taking it "to mean a medium in which electromagnetic actions take place in the absence of substances generally recognized as matter." "Of these actions — he added — the vibrations which constitute light are the most important:

---

[9]The second edition was also translated into French by the physicist and chemist Arthur Corvisy: *La théorie électrique moderne. Théorie électronique* (A. Hermann, Paris 1919).

26 *Einstein's Relativity in Great Britain*

studies of the properties of the aether usually resolve themselves into optical investigations." Actually, Campbell was not ready to accept for the ether any further roles. This is clear in Section 11 ("Aether and Energy") of Chapter XIV ("The Laws of Electromagnetism") of that first edition of *Modern Electrical Theory*, as well as in later works; for instance, in the paper entitled "The Aether," which he published in 1910.[10] There we find paragraphs like the following one (Campbell 1910: 181–182):

> "No doubt much of the dissatisfaction with 'the aether' is based on the recent theories of the atomic nature of radiation and on the proof that the principle of relativity is an adequate foundation for electromagnetic theory, but it is clear that such theories do not provide either a sufficient or a necessary reason for abandoning the conception. Sir J.J. Thomson, the author of the earliest and most far-reaching atomic theory of radiation, devoted much of his presidential address before the British Association to a description of the properties of the aether, while on the other hand, I hope to show that a consideration of no ideas more novel than the elements of electrostatics may lead to grave doubts concerning the utility of that conception."

The last sentence shows that Campbell's arguments against the ether concept did not follow only from his positivistic outlook, or from his adherence, posterior to 1907 to Einstein's special relativity theory.

III

Special Relativity in Britain: First Notices

1.8. The First Written Evidences of British Knowing the Existence of Einstein's Special Theory of Relativity

FitzGerald, Larmor, Poynting, Campbell (before 1905), the cases I have presented, belong to what one can denominate "the prehistory of

---

[10]This paper was the basis for the appendix on the ether included in the second edition of *Modern Electrical Theory* (1913).

relativity," neither of them could have known of Einstein's 1905 article, several years in the future. Thus, a question arises: when, and who was the first British which has left proof that he knew about Einstein's contribution. As far as I know, such merit has been awarded to George Augustus Schott (1868–1937), professor of Applied Mathematics in Aberystwyth from 1910 till his retirement in 1929, of whom I will say more in another section. In a paper Schott published in 1907, "On the Radiation from Moving Systems of Electrons, and on the Spectrum of Canal Rays", we found a note, number 2, which shows that he already knew Einstein's paper of 1905. Thus, he stated (Schott 1907a: 687):

> "It is incorrect to call (u', v', w') [defined on p. 659 of the paper] the velocity of the system [...]; the expression for this velocity has been given by Einstein (*Ann. Phys.* 322, p. 916). The results of this paper are, however, not appreciably affected thereby."

Therefore, by 1907 careful readers of the *Philosophical Magazine* could find Einstein's name referred too.

However, usually that merit has been credited to another scientist, George Frederick Charles Searle (1864–1954), a Cambridge physicist who carried on his professional career at the Cavendish Laboratory under J.J. Thomson.[11] Actually, in his autobiographical book, *Recollections and Reflections*, Thomson (1936: 114–115) said of him, remarks that at the same time add to our knowledge of how physics was taught then in Cambridge:

> "Mr. G. F. C. Searle, F.R.S., who first began to demonstrate [that is, to occupy the post of demonstrator] in 1891, has done more than anyone else for the teaching of practical physics at Cambridge. It was only in the year 1935 that, owing to the age limit, he gave it up after forty-four years' uninterrupted service, teaching all the time the largest class — that for students who are going in for Part I of the Natural Sciences

---

[11] Though he did not say explicitly that Searle was the first of knowing of Einstein's special theory of relativity — he talks only of "the Cambridge reception of Einstein's special theory of relativity" — in his splendid book, *Masters of Theory*, Andrew Warwick (2003) only studied Searle, not mentioning Scott in the book.

28 *Einstein's Relativity in Great Britain*

Tripos. Nothing approaching such a long tenure at this of the office of Demonstrator has ever come to my knowledge. In general the zeal for demonstrating fades away after a few years: the work is hard and may be monotonous, but Searle is as keen now as he was forty-four years ago; indeed it was only the other day (1935) that he invented a new and very beautiful experiment for his course. He has a real enthusiasm for the subject; he is not content with merely teaching it. He has taken infinite pains and thought to improve the course, and replace the experiments by others which have a greater educational value. His experiments are never commonplace: they have always a freshness and elegance which makes one want to do them. After he has got the idea, he works away at the experiment until he has found the form which gives the most accurate results, and so is suitable for training the students in accuracy of measurement. [...] He then makes a fair copy in his own hand, and this is placed in the laboratory for the use of the students. He must have made, I should think, many more than a hundred of these. Since for more than forty years practically every Cambridge man taking physics has been a pupil of Dr. Searle, the influence he has exerted on the teaching of practical work in physics must have been comparable with that exerted by Routh on the teaching of mathematics."

In 1900, Searle was appointed to a new Lectureship, at the same time that C.T.R. Wilson, who in 1925 was elected to the Jacksonian Professorship. Besides his excellent work as a teacher, which led to books like *Experimental Elasticity* (1908), *Experimental Harmonic Motion. A Manual for the Laboratory* (1915, 1922, 2nd edition), *Experimental Optics* (1925, 1935, 2nd edition), and *Experimental Physics* (1934), Searle worked on the dependence on the velocity of the electromagnetic mass, an interest which must have led him to Einstein's special theory of relativity. Such interest, indeed his knowledge on Einstein's work, seems to be due to the connections which he had established with other researchers interested in the problems related to the electron's variation of mass. One of them was the German physicist Alfred Heinrich Bucherer (1863–1927), who had published rather extensively on those questions, and with whom Searle, who was a regular visitor to Germany and was acquainted with several German physicists, was in contact: while attending the 80th meeting of the German Naturforscher versammlung held in September 1908, where he presented two papers, Searle was a guest of Bucherer, who

The Reception of Special Relativity in Great Britain 29

the previous year had published an article in the English *Philosophical Magazine*, entitled "On a new Principle of Relativity in Electromagnetism" (Bucherer 1907), a fact that perhaps facilitated the relation between both.[12] As Warwich (2003: 399) has explained, "Writing to his friend Oliver Heaviside in early March 1909, Searle [...] confided that he had 'no idea' what the 'principle of relativity' was." To remedy such ignorance, Bucherer wrote to Einstein asking to send Searle his 1905 paper, which he received. And on 20 May 1909, Searle wrote Einstein (CPAE 1993: 190–191):

"Dear Sir:

I am sorry that I have so long delayed to write you for sending me — at the request of Dr. Bucherer — a copy of your paper on the principle of relativity. When the paper came to me I was rather tired with my work. Then came a holiday. But as soon as I returned from the holiday I fell ill and have been unwell up to the present time. I am now recovering and hope to be quite well in a few days. I had hoped to make a careful study of the paper before writing to you, but I have not been able to do so.

I have not been able so far to gain any really clear idea as to the principles involved or as to their meaning and those to whom I have spoken in England about the subject seem to have the same feeling.

I think it would help us if you were to write a short account of the subject which could be translated and published in some English journal — the *Phil. Mag.* perhaps. If I could give any assistance in the matter I should be glad to do son, but I must say that I am not an expert in German.

I send a few papers which I hope you will accept. They are on my own lines and have very little reference to the work of others. This is mainly because my teaching work prevents me from reading very much of what is written by others.

I have been in Berne several times. My wife and I were there for one night in Sep. 1907. For the first time I saw the Alps perfectly clear. I thought they looked almost more beautiful at that distance than from Mürren or Grindelwald.

---

[12] In an article published in 1906, Bucherer (1906) had been the first to use the expression "Einsteinian relativity theory" ("*Einsteinsche Relativitätstheorie*").

30 *Einstein's Relativity in Great Britain*

If I find myself at Berne again, I hope I may have the pleasure of meeting you.

I as recently at Torquay and had several conversations with Dr. Oliver Heaviside on electrical matters. I think he would be interested to have a copy of your paper. His address is Homefield Lower Warberry Road Torquay. With renewed thanks for your paper I remain yours truly,

G. F. C. Searle."

1.9. George A. Schott

The works of the already mentioned George Augustus Schott show many of the principal characteristics of the introduction and development of special relativity in Great Britain.[13] Especially interesting in this sense is Schott's preface to his great monograph, *Electromagnetic Radiation* (Schott 1912), initially an Adams Prize essay at the University of Cambridge in 1909.[14]

Almost at the end of the preface Schott (1912: xv) pointed out his great debt to J. Larmor and H.A. Lorentz, "whose writings have furnished the theoretical foundation of this essay," as well as — and this is important — to his former teacher, J.J. Thomson, "to whose paper on Cathode Rays this investigation owes its inception."

I have said that this is important because one of my points is that it is not possible to understand completely the process of introduction and development of special relativity in Britain without taking into account: (i) the discoveries and researches which were being carried out in the realm of "atomic physics", and (ii) the efforts which were undertaken for explaining those new results of atomic physics within the framework of electrodynamics. Schott (1912: v–vi) was very clear in this regard. Thus, he wrote:

"In consequence of the discoveries of the last few years in the fields of radioactivity, vacuum-tube phenomena, magnetism and radiation, a

---

[13] For details of Schott's career see Conway (1936–1938).

[14] The original title of the essay was *The Radiation from Electric Systems or Ions in Accelerated Motion and the Mechanical Reactions on their Motion which arise from it.*

The Reception of Special Relativity in Great Britain 31

great need has arisen for a comprehensive Electron Theory of Matter, which shall systematize the results already achieved, as well as serve as a guide in future researches [....] a beginning as has been made by J.J. Thomson in his well-known paper on the Structure of the Atom and in his books on Electricity and Matter and the Corpuscular Theory of Matter."[15]

The problem was, however, that Thomson's investigations had been carried out under the restriction that the electrons in the atomic model move with velocities so small, compared with the velocity of light, that they can be treated like the particles of ordinary mechanics. Schott, as well as many others, knew, however, that this was not true. By that time it was well-known that moving electric charges do not behave like the particles of ordinary mechanics; for instance, their mass varies with their speed, and they generate a magnetic field which reacts to their motion in various ways (*radiation-reaction*).[16] When the speeds of the charges are small compared with that of light, "these efforts" — Schott pointed out — "are small but not negligible." For example, Walther Ritz's then well-known theory of the production of spectrum series rested on the effect of a magnetic field in the atom on the motion of electrons. There was, moreover, the problem of the assumption of the smallness of the velocity of light. According to Schott (1912: vi):

"it is by no means certain that all the electrons inside the atom are moving with speeds small compared with that of light; we know that β-particles are expelled from comparatively stable atoms, like that of Radium, with speeds differing from that of light by only 2%. The kinetic energy of such a β-particle amounts to three millionths of an erg, which is five times the mutual electrostatic energy of two negative electrons in contact. It is not easy to imagine an arrangement of negative and positive charges in equilibrium, or in slow stationary motion, which shall be

---

[15] In a paper he published in 1907, Schott (1907b: 189) defined what he understood by "Electron Theory of Matter;" namely "any theory which assumes matter to consist of electrical charges, acting upon each other with electromagnetic forces only".

[16] This was one of the main topics of Schott's investigations in the *Electromagnetic Radiation*.

32 *Einstein's Relativity in Great Britain*

sufficiently permanent and stable to serve as a model of the Radium atom, and at the same time capable of setting free sufficient potential energy to supply the kinetic energy of a β-particle and also overcome the attraction of the positive charges. If on the other hand we suppose the β-particle to be already moving inside the Radium atom with a speed comparable with that of light this difficulty does not arise."

Actually, Schott's endeavour since at least 1906 was precisely that one: To explain "atomic phenomena" by means of "classical physics," electrodynamics in particular. It was in that way that he came across problems related to special relativity; but it was, as it were, a sort of offshoot from his specific topic of research: An electrodynamical theory of atomic phenomena (electron theory of matter). My point is that, in this sense, Schott is just one example of a way to approach the "relativity worldview," which was common to many of the physicists, British or not, who by that time were trying to understand the large number of new phenomena which were being discovered. As we shall see, the ether concept was not necessarily one of the premises of such approach.

The preface of *Electromagnetic Radiation* is also worth to quote from, as in it Schott (1912: vii) mentions the "postulate of Relativity:"

"Let us now consider the fundamental assumptions on which the present investigation is based. [....] In choosing the fundamental assumptions I have throughout aimed at securing the greatest generality consistent with throughly well establish experimental results. Additional assumptions have only been introduced when further progress seemed impossible without them, or when a comparison of results already obtained with experiment clearly indicated that further restrictions were desirable. For these reasons *I have refrained from making any use, either of the Postulate of Relativity, or the Aether Hypothesis, which by some are regarded as inconsistent with each other.* Some of the results obtained in this essay are consistent with the Postulate of Relativity; others cannot be reconciled with it, at least when it is used in the strictest possible sense, and find their natural explanation in terms of the aether. All however are deduced quite independently of either hypothesis. *It does not appear to me that either of the new theories is so well established and so generally accepted yet as to be properly made the basis of an investigation in which the utmost generality is aimed at* [emphasis added]."

One of the problems to which Schott was referring dealt with an objection raised by Max Abraham in the second edition (1908) of the second volume ("Der Elektromagnetische Theorie der Strahlung") of his famous textbook *Theorie der Elektrizitä t.* There, Abraham pointed out that, according to the theory of relativity, the mass, m_o, and the electric energy, W_o, of a slowly moving electron satisfy the relation $m_o = W_o/c^2$, while the mass of the Lorentz electron (not a *point* electron) is 4/3 of the amount determined by this equation. Schott was able to show — see equations (350), Section 229, of *Electromagnetic Radiation* — that even for a symmetric electron m_o always exceeds W_o/c^2, whatever the configuration of the electron may be when it is at rest. He also found that an extended electron, whatever the distribution of its charge may be, cannot exist unless it is subjected to a suitable pressure on its outer surface. According to him, this meant that (p. xiv): "We must either postulate the existence of some external medium which shall produce the required surface pressure, or admit that the elements of charge of the electron exert on each other actions at a distance, which are not electromagnetic and follow quite different laws."

To Schott (1912: xiv), the "first hypothesis, that of an external medium, appears to be the more reasonable of the two and amounts to admitting the existence of the electromagnetic aether." His conclusion is clear:

"Thus it appears that the acceptance of the Postulate of Relativity in its strictest form almost necessitates the adoption of the hypothesis of the extended charge, while the hypothesis of the extended charge leads naturally to the adoption of the aether hypothesis."

This was, however, only one aspect of the "controversy *Postulate of Relativity-Aether Hypothesis,*" and the preface makes it clear that Schott did not think the controversy was settled.[17] How little ether-dependent was

---

[17] Actually, it seems that Schott accepted the special theory of relativity. In the obituary already referred to, Conway (1936–1938: 453) wrote: "The theory of relativity was accepted by him, although mathematical difficulties in his latest papers [in the thirties] caused him to use the conception of a rigid sphere." Let me add, however, that 1 think that Conway's statement can be criticised on different grounds.

34 *Einstein's Relativity in Great Britain*

Schott's approach can be seen, for instance, when, referring to Maxwell-Hertz's electromagnetic equations, he writes (Schott 1912: vii): "These equations already imply the existence of a system of axes to which they are referred; *it is immaterial for our purpose whether these axes be regarded as fixed in space, relative to a fixed aether, or as only fixed relative to the observer*" [emphasis added].

With Schott, we have, therefore, one example of a British physicist who carried on high-level research related to relativity, and who, nevertheless, was not *a priori* committed to the ether concept.

1.10. Norman R. Campbell (2)

Now let us return to the case of Norman Robert Campbell to show, as in Schott's case, but in a probably clearer manner, that the ether concept was not necessarily behind all the presentations and researches which were being carried out in the field of electrodynamics. We have already seen that Campbell was a fierce critic of the ether concept, but we have not yet discussed the role played by Einstein's special theory of relativity in the two editions of his book, *Modern Electrical Theory*.

As far as special relativity is concerned we have that, in the first edition (Campbell 1907) of *Modern Electrical Theory*, Einstein's name or ideas do not appear. He referred only to Bucherer, Lorentz, Abraham or Larmor, among others. This fact suggests that, in contrast with the case of Schott, by 1907 Campbell was not acquainted with Einstein's 1905 paper. In fact, it seems that by that time Campbell was rather close to *some* points of the "Electromagnetic View of Nature", though a rather peculiar version, as he was not sympathetic to the ether concept. Thus, he wrote (Campbell 1907: 302): "Now we have seen that it is probable that all the properties of material bodies — chemical, mechanical, optical, thermal and the rest — can be reduced to electrical properties."[18]

In this sense, it is important to point out that Campbell (1907: 303) thought that Michelson and Morley's experiment supports the electromagnetic worldview:

---

[18] This is not to say, however, that he did not see any problems with such a worldview. He realized, for instance, that there were some difficulties connected with aberration.

"From this point of view, this negative result [Michelson and Morley's] proves that the optical properties of matter (on which the experiment is based) are electrical in origin — a conclusion that nobody doubts nowdays. Accordingly the Lorentz-Fitzgerald hypothesis has met with general acceptance, and it has been concluded that the difficulties which were raised by the Michelson-Morley experiment have been solved satisfactorily."

We see, therefore, that Campbell's point of departure was electrodynamics (an electrodynamics free from the ether, a concept that he did not accept). The question is: did he "arrive" at special relativity, and if so, when? Indeed, he did. By 1913, when the second edition of *Modern Electrical Theory* appeared, Campbell (1913) already knew about Einstein's special relativity theory. The new preface is quite clear on this point: "This volume, nominally, a second edition, is really a new book. [...] The Principle of Relativity and Stark's work on atomic structure have altered Part III completely."

But Campbell not only knew *about* special relativity, he also understood it quite well. "This last chapter — he wrote (Campbell 1913: 351) — will be devoted to a consideration of certain problems which in the past have been the subject of much discussion among physicists. These problems have little direct connection with those which have concerned us hitherto and our study of modem electrical theory would be logically complete with only the barest reference to them." That is, Campbell understood what is one of the fundamental ideas of the special theory of relativity; namely, its basic independence from electrodynamics, or, in other words, that it is a *kinematics* which ought to be applied to any interaction (to the electromagnetic one too, of course). Most probably, his methodological opinions, which led Campbell to reject the ether, would help him in understanding and accepting Einstein's relativity. In this sense, it is worth quoting from appendix I ("The Aether") to the second edition of *Modern Electrical Theory* (1913: 388)[19]:

"very shortly Einstein showed that the rules given by this complicated theory [i.e., Lorentz's and Fitzgerald's] for deducing the relation

---

[19] This appendix reproduces an article of Campbell (1910).

36 *Einstein's Relativity in Great Britain*

between the laws observed by two observers who are in relative motion are exactly the same as those given by the Principle of Relativity; these rules, of course, involve only the relative velocity of the two observers and not their 'velocity relative to the aether.' Moreover, the form in which those rules are expressed by the Principle of Relativity is much simpler and more convenient than that in which they are given by the aether theory combined with the Lorentz–Fitzgerald hypothesis

1.11. Ebenezer Cunningham

One of the first references made in Great Britain to Einstein's 1905 contribution was due to a former Senior Wrangler (1902) and winner of the Smith Prize in 1904, Ebenezer Cunningham (1881–1977), whose career took him to the University of Liverpool (1904), University College London (1907), where he worked with the biostastician Karl Pearson, and finally to his *ala mater*, St John's College, Cambridge, where he remained the rest of his career as lecturer.

In a paper he published in October 1907 issue of the *Philosophical Magazine*, "On the Electromagnetic Mass of a Moving Electron," Cunningham (1907) joined the, at the time, important discussion of the electromagnetic mass of a moving electron. Specifically, he opposed Max Abraham's (1905: 205) objection to Lorentz's conception of an electron as having, at rest, a spherical shape, but in motion the shape of an oblate spheroid. "The present paper — Cunningham (1907: 539) wrote — reconsiders Abraham's discussion and comes to the conclusion that the objection is not valid." The reasons why he arrived at such a conclusion were based on the fact that "it has been proved that Maxwell's equations represent equally well the sequence of electromagnetic phenomena relative to a set of axes moving relative to the aether, as relative to a set of axes fixed in the aether."

The proof in question was none other than the substitution of Galileo transformations between inertial frames of reference for what is now called Lorentz transformations, whose expression Cunningham duly reproduces in his paper, adding that the "above transformation renders the electromagnetic of a system independent of a uniform translation of the whole system through the aether."

The Reception of Special Relativity in Great Britain 37

For my purposes here, it is important to point out that, while dealing with that electromagnetic problem, Cunningham showed his knowledge of Einstein's contribution, the special theory of relativity. "The transformation in question — he wrote in a footnote (Cunningham 1907: 539) — is given by Einstein in a paper in the *Annalen der Physik* [...]. It is in substance, the same as that given by Larmor in 'Aether and Matter,' chap. xi, though the correlation is only proved as far as the second power of v/c. Prof. Larmor tells me he has known for some time that it was exact. Vide also Lorentz, *Amsterdam Proceedings, 1903–1904*." However, "knowledge of the *existence* of" does not necessarily entail "knowledge of the *meaning* of." Thus, taken by itself, Cunningham's 1907 paper does not offer any evidence that he understood that Einstein's theory was different from Larmor's or Lorentz's, in the sense that his special theory of relativity transcended electromagnetism. Quite on the contrary, the terminology, together with the exclusive electromagnetic framework and the persistent use of the expression "motion through the aether" does strongly suggest that Cunningham was still "on Larmor-Lorentz's side;" that is, that he believed that Lorentz's transformations were just a sort of theorem, or property, of Maxwell-Lorentz's electrodynamics.

The point is, nevertheless, that Cunningham's early relationship with special relativity provides us with another manifestation — different from Schott's or Campbell's cases — of the close connection existent between the introduction of Einstein's theory in Great Britain and the development of electrodynamics.

There is another fact, related to Cunningham's paper, which must be mentioned here. In his readiness to acknowledge Einstein's 1905 contribution, Cunningham behaved in a somewhat different manner than some of his German colleagues. Actually, we find in the pages of the *Philosophical Magazine* one illustration of this fact. In its April 1907 issue the then *Privatdozent* at Bonn University, Alfred Heinrich Bucherer, published a paper entitled "On a New Principle of Relativity in Electromagnetism." In that paper Bucherer (1907), "guided by the feeling that the *form* of the Maxwellian equations must correspond somehow to the true laws of electromagnetism," attempted a new interpretation of these equations that would "harmonize with facts." Although the *spirit* (not the structure, or even the details) of Bucherer's ideas was clearly in agreement with

38   *Einstein's Relativity in Great Britain*

Einstein's, he nevertheless did not mention or refer to Einstein's name. He only said that (Bucherer 1907: 413–414):

> "It is needless to dwell on the serious difficulties which the Maxwellian theory has encountered by the well established experimental fact that terrestrial optics is not influenced by the earth's motion. The endeavours of some distinguished physicists, notably of H.A. Lorentz, to modify the Maxwellian theory in su eh a manner as to eliminate the effects of translatory motion have admittedly failed."

It was only when some of his contentions were criticized by Cunningham, in the above cited paper (where Einstein's name was mentioned), that Bucherer spoke of the "Lorentz-Einstein Principle."

I will return to Cunningham in another section, in which I will talk about a monograph he dedicated to special relativity.

1.12. Harry Bateman

With Harry Bateman (1882–1946), a Manchester-born mathematical physicist who became fellow of Trinity College, Cambridge, and Reader in Mathematical Physics at the University of Manchester, before he settled (1910) in the U.S.A., we have a different "electrodynamical approach" to relativity. It is, in this sense, important to pay attention to him, inasmuch as he will help us to realize that there was a broad range of approaches to special relativity based on Maxwell's electrodynamics.

The essence of Bateman's approach lies in its geometrical character. "Recent theoretical researches in electromagnetism — Bateman (1910: 623) wrote in one of his papers — indicate that the science of electromagnetism is closely connected with the geometry of spheres." Taking into account this fact, we should not be surprised to learn that Bateman had Hermann Minkowski's four-dimensional ideas in great consideration (we shall consider later on the role played by Minkowski's point of view in the introduction of relativity in Great Britain). Thus, we see that in the paper just referred to, Minkowski's name appears alongside those of Lorentz, Larmor, and Einstein. The following quotation, from the same paper (Bateman 1910: 624) is truly Minkowskian: "The group of *Lorentzian transformations* for which the electron equations are covariant is then represented by the group of transformations of rectangular axes in the space of four dimensions."

Another aspect of Bateman's interests on this regard, are the papers he devoted to the conformal transformation. In 1909, he had published a paper entitled "The Conformal Transformations of a Space of Four Dimensions and Their Applications to Geometrical Optics" (Bateman 1909), in which he showed that Maxwell's equations are covariant not only under the Lorentz group but also under the more general fifteen-parameter conformal group. As a matter of fact, this discovery was made independently by Cunningham. (The title of the paper which Cunningham published in 1910, "The Principle of Relativity in Electrodynamics and an Extension Thereof", is significant in as much as it shows that Bateman's and Cunningham's approaches could contribute to the development of special relativity along routes not necessarily connected with, for instance, the ether concept. As a matter of fact, Cunningham acknowledged in a footnote that [Cunningham 1910] that "This paper contains in an abbreviated form the chief parts of the work contributed by the Author to a joint paper by Mr. Bateman and himself read at the meeting [of the London Mathematical Society] held on 11 February 1909, and also the work of the paper by the author read at the meeting held on 11 March 1909".).

What has been said about Bateman should not be taken as if he were completely foreign to the problems aroused by the ether concept. A few quotations from the book, *The Mathematical Analysis of Electrical and Optical Wave-Motion,* (Bateman 1915: 2) will serve our purposes in that sense:

> "Some of the modern writers on the theory of relativity maintain that the introduction of the idea of an aether is unnecessary and misleading. Their criticisms are directed chiefly against the popular conception of the aether as a kind of fluid or elastic solid which can be regarded as practically stationary while material and electrified particles move through it. This idea has been very helpful as it presents us with a vivid picture of the processes which may be supposed to take place, it also has the advantage that with its aid we can attach a meaning to the term absolute motion, but here in lies its weakness. Larmor, Lorentz and Einstein

40   *Einstein's Relativity in Great Britain*

have shown, in fact, that the differential equations of the electron theory admit of a group of transformations which can be interpreted to mean that there is no such thing as absolute motion.

If this be admitted, the popular idea of the aether must be regarded as incorrect, and so if we wish to retain the idea of a continuous medium to explain action at a distance we must frankly acknowledge that the simplest description we can give of the properties of our medium is that embodied in the differential equations [rot $\mathbf{H}$ = $(1/c)$ ∂ E/∂ t, rot $\mathbf{E}$ = $-(1/c)$ ∂ H/∂ t]."

This last sentence is, in some sense, not too different from Hertz's famous dictum: "Maxwell's theory is Maxwell's system of equations." It can be taken as implying that Bateman did not care much about the ether; a conclusion which is reinforced if one reads this book, a truly mathematical-physics monograph not much preoccupied with physical concepts. Still, the content of the only pages of *The Mathematical Analysis of Electrical and Optical Wave-Motion* explicitly devoted to a discussion of the concept of interaction suggests that Bateman (1915: 2–4) was inclined to favour explanations based on the ether concept, although, at the same time, leaving open other possibilities:

"If we abandon the idea of a continuous medium in the usual sense only two ways of explaining action at a distance readily suggest themselves. We may either think of the aether as a collection of tubes or filaments attached to the particles of matter as in the form of Faraday's theory which has been developed by Sir Joseph Thomson and N.R. Campbell; or we may suppose that some particle or entity which belonged to an active body at time t belongs to the body acted upon at a later time $t + \gamma$. From one point of view these two theories are the same, for if particles are continually emitted from an active body they will form a kind of thread attached to it. The first form of the theory is, however, more general than the second."

Bateman's mathematical outlook can be easily seen when he writes (p. 3): "At present we are unable to form a satisfactory picture of the processes that give rise to, or are represented by, the vectors E and H. We believe, however, that some points may be made clear by studying the properties of solutions of our differential equations." And his mathematical analysis of those equations led Bateman to three distinct theories of the universe, which may be described briefly as follows:

ETHER
- Continuous medium.
- Discontinuous medium consisting of a collection of tubes or filaments.
- Continuous medium.

MATTER
- Aggregates of discrete particles.
- An aggregate of discrete particles attached to tubes.
- An aggregate of discrete particles to which tubes are attached.

"The last theory — Bateman (1915: 3) pointed out — may be supposed to include that form of the emission theory of light in which small entities are projected from the particles of matter under certain circumstances and produce waves in the surrounding medium. This theory might be justly ascribed to Newton." However, it was to the first theory that most of Bateman's book (especially the first part) was dedicated. "The other theories — he added — have not yet received much attention but it is hoped that the analysis of Chapter VIII will lead to further developments so that a comparison can be made between the different theories. *It is likely that one theory will be enriched by the developments of another*" [emphasis mine].

1.13. James H. Jeans

There is no doubt that textbooks have great importance in the way new ideas are accepted or rejected, at the same time as they reflect the state of the art when they were written. If the book has been an influential one, then the historian has, by analysing its content, an excellent instrument to carry out his work. This is the case with one of the several books that the influential British physicist and astronomer James Hopwood Jeans (1877–1946): *The Mathematical Theory of Electricity and Magnetism*, an

42 *Einstein's Relativity in Great Britain*

excellent and widely-used textbook. Even as this book adds to the interest of its author — it was written by one of the most popular and influential physicists in Britain during the first decades of the XXth century — the fact that it is dedicated to electromagnetism helps in studying the changes undergone in its different editions as far as concerns the theory of relativity.

The first edition (Jeans 1908) of the book did not contain any discussion of relativity. It was only in the second edition (1911) that Jeans added, as he wrote in the preface, "two new chapters [...] on the Motion of Electrons and on the General Equations of the Electromagnetic Field," where some problems *related to* relativity were touched. It is important to point out, however, that Jeans's approach was fully within the ambit of electrodynamics; in this sense, it did not differ from, for instance, the first edition (1907) of Campbell's *Modern Electrical Theory*. So, he dealt firstly with the problem of the variation, with respect to velocity, of the kinetic energy of an electron moving with any velocity, referring to the experiments performed by Walther Kaufmann and Alfred Bucherer, as well as to their significance for the theories proposed by Abraham and Lorentz. After that, Jeans considered the question of the motion of an arbitrary system in equilibrium. The problem here was that when a material system moves, the electric field produced by its charges is different from the field when at rest; the difference between these fields must show itself in a system of forces which must act on the moving system and in some way modify its configuration. Jeans showed in which sense the equilibrium of the original system was maintained. Once this question was settled, he stated (Jeans 1911: 577): "Lorentz, to whom the development of this set of ideas is mainly due, and Einstein have shewn how [this result] may be extended to cover electromagnetic as well as electrostatic forces." This was the only reference to Einstein made by Jeans in 1911.

The last two problems discussed by Jeans in the second edition of his textbook were the "Lorentz-Fitzgerald contraction hypothesis" and the "Lorentz deformable electron." The beginning of the section dedicated to the Lorentz–Fitzgerald contraction is significant (Jeans 1911: 577).

"It is now natural (Jeans wrote to make the conjecture, commonly spoken of as the Lorentz-Fitzgerald hypothesis, that the system S when set in motion with a velocity U assumes the configuration of the system S', this latter being a configuration of equilibrium for the moving system. *Indeed, if we suppose all forces in the ether to be electrical in origin, this view* is *more than a conjecture; it becomes inevitable.* Put in the simplest form it asserts that any system when set in motion with uniform velocity V is contracted, relatively to its dimensions when at rest, in the ratio $(1 - V^2/c^2)^{1/2}$ in the direction of its motion (emphasis mine)."

That is, Jeans, like Campbell in 1907, viewed the Lorentz–Fitzgerald hypothesis as part of the electromagnetic worldview; there is nothing in his 1911 presentation revealing that he understood such hypothesis in the way Einstein did in his seminal paper of 1905.

Almost a decade later, Arthur Eddington (1920) felt obliged to clarify the issue at debate:

"When we consider the matter carefully, it is not so surprising after all. The size and shape of the material apparatus is maintained by the forces of cohesion, which are presumably, of an electrical nature, and have their seat in the aether. It will not be a matter of indifference how the aether is streaming past, and there will be a readjustment of the molecules when the flow is altered.

But what does seem surprising is that the readjustment of size should just hide the effect we were hoping to find. It looks almost like a conspiracy. There have been three or four other experiments tried since then — of more technical kinds — all in the hope of detecting our motion through the aether; but they have all been defeated by the same kind of conspiracy. And to turn from experiment to theory, we find the same kind of conspiracy even in our mathematical equations. Perhaps you know sometimes, just as you are hoping to get an important result, the equations evade you, and, instead of telling you what x is, they announce the solemn but irritating truth that $0 = 0$. It is somewhat like this in our electromagnetic theory: the velocity through the aether appears abundantly in the formulae, but, whenever we try to run it down, it drops out, and refuses to be equal to anything in particular.

In consequence of these conspiracies, a great generalisation has been put forward, known as the restricted Principle of Relativity. *It is impossible by any conceivable experiment to detect the velocity of a system through the aether.* That is to say, the conspiracy is a general

44 *Einstein's Relativity in Great Britain*

one; and, in fact, it is in the nature of things that this motion is undetectable.

We make this generalisation and build a branch of science on it, in the same way as we make the generalisation that it is impossible to construct a perpetual-motion machine, and build the science of thermodynamics upon it. The case for the Principle of Relativity seems to me as strong as the case for the denial of the possibility of perpetual-motion machines."

The conclusions of the last section of the book, dedicated to the Lorentz deformable electron, reinforced Jeans's belief in the electromagnetic view of nature. Commenting on the agreement between Bucherer's experiments and Lorentz's formulae, Jeans (1911: 579) stated that Bucherer's experiments seemed to lead to the conclusion (among two other possible ones) that "the mass of the electron is purely electromagnetic in its nature." This is the last sentence of Jeans's book.

The real changes in Jeans's textbook, as far as special relativity is concerned, arrived with the fourth edition (1920). In the preface, dated December 1919 (recall that 1919 is the year of the British expedition, led by Dyson and Eddington, to observe the solar eclipse of May 29), Jeans made clear which were the novelties: "It will be found that the main changes in the fourth edition consist in a rearrangement of the later chapters and the addition of a whole new chapter on the Theory of Relativity." Looking at the actual content of this new edition, we find that, although Jeans was quite slow in opening the pages of his book to Einstein's theory, he, at last, understood it now quite well. Thus, the manner in which he introduces the "relativity-condition" is clearly a generalization that superseded specific interactions like the electromagnetic one.[20] Furthermore, he realized the consequences that Einstein's point of view had for the ether (Jeans 1920: 619):

"The hypothesis that there is an ether may give a possible explanation of the phenomena, but the hypothesis that there is no ether provides an equally possible and very much simpler explanation. [...] If the observed

---

[20] Jeans's formulation is quite clear: "Systems of equations or natural laws which are such as to make it impossible to determine absolute motion may be said to satisfy the 'Relativity-condition'" (Jeans 1920: 597).

constant velocity of light is simply the constant velocity of propagation through an ethereal medium, it would seem to follow that each observer must carry a complete ether about with him. This at least robs the ether of the greater parts of its reality. [...] Considerations such as we have mentioned do not prove in strictness that light cannot be propagated through an ether; what they prove is that if an ether exists, it must be something very different from the absolutely objective ether imagined by Maxwell and Faraday.

Ambiguous as the last paragraph may seem (it was not so easy to escape from one's upbringing), Jeans's stand about the ether was quite clear. He considered it as an unnecessary device. Indeed, he even gave a rather original argument against that elusive concept (Jeans 1920: 620):

"If an ether existed it would provide a fixed set of axes relative to which all positions and velocities could be measured. To account for the result of the Michelson-Morley equipment, it would be necessary to postulate a real shrinkage of all bodies moving through the ether. This shrinkage could not be detected by mechanical means, for a measuring rod would shrink in precisely the same ratio as the body to be measured, but it could be detected by gravitational means unless every gravitational field of force shrunk in just a way as to conceal the shrinkage of matter.

At this point, Jeans argued that if the gravitational field were not to shrink then a geoid, or surface of mean sea-level on the earth, might be a gravitational equipotential for some velocity through the ether, but could not remain an equipotential as the earth's velocity through the ether changes on account of the description of its orbit. That is, one would expect to observe seasonal and even daily surgings resulting from the earth's motion *through the ether*. However, no such events are observed, "whence — Jeans continued — it seems natural to suppose that gravitation al so must conform to the relativity-condition," a conclusion that, he notes, was supported by the fact that "Einstein's relativity theory of gravitation has received confirmation." Consequently, it was almost inescapable to conclude that (Jeans 1920: 621):

"If, then, we continue to believe in the existence of an ether, we have also to believe not only that all electromagnetic phenomena are in a

46   *Einstein's Relativity in Great Britain*

conspiracy to conceal from us the spread of our motion through the ether, but also that gravitational phenomena, which as far as is known have nothing to do with the ether, are parties to the same conspiracy. The simpler view seems to be that there is no ether. If we accept this view, there is no conspiracy of concealment for the simple reason that there is no longer anything to conceal."

1.14. Another Type of Approach: Alfred A. Robb

It has been my purpose in the preceding sections to show that there existed a rather large number of different approaches in Great Britain towards the sort of problems that characterize special relativity. It is true, however, that all the approaches reviewed so far were, in one way or another, closely related to electrodynamics. The case of Alfred A. Robb (1873–1936), a former student in J.J. Thomson's Cavendish Laboratory, and F.R.S. since 1921, is completely different. As a matter of fact, Robb's contributions inaugurated a new phase in the understanding of relativity, a "style" which, it true, lacked of influence during the first decades of the special and general theories of relativity, in contrast with the strength of the influence it had later on, with the pioneer works of Roger Penrose and Stepehn Hawking.[21]

Thought he was not as famous as other physicists or mathematicians who dealt with problems related to special relativity, Robb was not an unknown personage. According to A. Vibert Douglas (1956: 99), Eddington's biographer, "[L.A.] Pars recall[ed] frequently seeing [Eddington] strolling along Jesus Lane after the service to call on A.A. Robb, a kindred spirit both because of his deep grasp of the problems of time and of cosmic order, and by reason of his gift for whimsical rhyming". And in the interview of which he was protagonist as part of the "Sources of for the Quantum Physics" project, the Dutch Adriaan D. Fokker, a former student of H.A. Lorentz, recalled that during 1913–1914 he "was much struck by the papers of Robb, who wrote about the theory of Einstein. In fact, he wrote an axiomatic theory for Einstein, avoiding the word relativity." Actually, during his stay in England (1914) as a post-doctoral student, Fokker went to Cambridge only to make the

---

[21] Robb's work has been studied, from a mathematical point of view, by Briginshaw (1979).

acquaintance of Dirac and Robb. Later on, he would write, in 1929, the first book dedicated to relativity to appear in Dutch.

It is also perhaps more than a coincidence that at the time that Robb was completing the first version of his theory, to appear as *Optical Geometry of Motion: a New View of the Theory of Relativity* (Robb 1911), Whitehead and Russell were publishing (1910–1913) their famous work, *Principia Mathematica* (1910–1913) that aimed at reducing mathematics — arithmetic in particular — to the principles of logic, also a landmark in what could be termed the "axiomatical approach" (to mathematics in this case). Thus, as early as 1911 Robb contributed to the widening of the modes of facing — or of understanding — the theory of special relativity, in Great Britain as well as in all the other countries.

After *Optical Geometry of Motion: a New View of the Theory of Relativity*, he put forward his views of space and time in two books, closely related: *A Theory of Time and Space* (Robb 1914), and *Geometry of Time and Space* (Robb 1936).

The origin and development of Robb's ideas were explained by himself in the preface of his book *Geometry of Time and Space* (Robb 1936). There, he wrote (Robb 1926: v–vii):

"The present volume is essentially a second edition of one which was published by the author in 1914 under the title: *A Theory of Time and Space*. An alteration of the title has been made, since it was considered that the word geometry conveyed a somewhat better idea of the nature of the contents of the book than did the word theory.

The first edition was going through the press at the time of the outbreak of the war, so that its publication took place under very unfavourable circumstances. The present volume differs from its predecessor in several respects. The Introduction has been re-written and extended; while the proofs of a number of theorems, which were rather lengthy, have been curtailed and simplified.

A considerable amount of new matter has also been introduced, making the book more self-contained and complete.

The demonstrations have all been carried out as deductions from certain postulates expressed in terms of the relations of after and before; so that the whole work may be regarded as a demonstration of the fundamental character of these relations in Time–Space theory.

48   *Einstein's Relativity in Great Britain*

So far as I am aware the book, in its original form, was the first of its kind to be written, and a brief account of its origin may be of interest. At the meeting of the British Association held at Belfast in 1902 Lord Rayleigh gave a paper entitled: "Does Motion through the Ether cause double Refraction?," in which he described certain experiments which he had carried out with the object of testing this matter, and which seemed to indicate that the answer was in the negative.

1 remember that he inquired of Professor Larmor, who was present on this occasion, whether, from his theory, he would expect double refraction to be produced in this way. Professor Larmor replied that he would not, and, in the discussion which followed considerable surprise was expressed that, in any attempt to detect motion through the aether, things seemed to conspire together so as to give null results. The impression which this discussion made upon me was, that in order properly to understand the matter, it would be necessary to make some sort of analysis of one's ideas concerning equality of lengths, etc., and I decided that, at some future time, I should attempt to carry this out. I am not quite certain that I had not some idea of the sort prior to this meeting, but, in any case, the inspiration carne from Professor Larmor, either then, or on some previous occasion while attending his lectures.

Some years later I attempted to carry out this scheme, and, while doing so, I heard for the first time of Einstein's work.

I may say that, from the first, I felt dissatisfied with his approach to the subject, and I decided to continue my own efforts to find a suitable basis for a theory.

The first work which I published on the subject was a pamphlet which appeared in 1911 entitled: *Optical Geometry of Motion: A New View of the Theory of Relativity.*

This pamphlet was of an exploratory character and did not profess to give a complete logical analysis of the subject; but nevertheless, although bearing a very different aspect, it contained some of the germs of my later work. It was, in fact, an attempt to describe Time–Space relations without making any assumptions as to the simultaneity of events at different places. Later on, the idea of *Conical order* occurred to me, in which instants at different places are regarded as definitely distinct; so that there is no such simultaneity.

As it was evident that a thorough working out of this idea would entail a great deal of labour, I published, in 1913, a short preliminary account of it under the title: *A Theory of Time and the book and Space.*

In 1914, as above mentioned, I published a book bearing the same title, of which the present volume is a second edition.

The working out of a scientific theory in the form of a sequence of propositions, such as was done by Euclid, Newton and others, seems largely to have gone out of vogue in these latter days and I consider that this is rather regrettable.

No doubt, in doing exploratory work, other methods are permissible and necessary, but I think that the incorporation of the more fundamental parts of a theory in a sequence of propositions should always be kept in view, since, in this way, one is able to see much more readily what are our primary assumptions, and one is able to fall back upon these in cases of difficulty."

In the "Introduction" to *A Theory of Time and Space* he was more explicit concerning Einstein's contribution. What he said then is worth to quote as it revealed much of how at least some British scientists thought about the special theory of relativity (Robb 1914: 1):

"Under the name of 'The Theory of Relativity' this subject has been much under discussion, but it is still in a condition of considerable obscurity.

Although generally associated with the names of Einstein and Minkowski, the really essential physical considerations underlying the theories are due to Larmor and Lorentz."

Robb's most fundamental idea was that the axioms of geometry are the formal expression of certain optical facts which are at the basis of special relativity, a theory in which, according to Einstein, the velocity of light plays a most important role, a "natural" idea for someone willing to set up a theory consistent with such an idea to choose the axiomatic approach. In fact, Robb has been nicknamed "the Euclid of Relativity." His 21 axioms and 206 theorems form what is probably the most complete and rigorous exposition of what the special theory of relativity has to say about space and time, as expressed in terms of a single basic concept, that of "conic order." Stated in a few words, it can be said that he was the first to work out the idea that the geometry of the Minkowski's space-time of special relativity can be developed on a purely causal basis: he derived the

50　*Einstein's Relativity in Great Britain*

entire geometry of Minkowski's space-time from an axiom system with only two primitives (undefined predicates), namely "*x* is an *element*," and "the element *x* is *after* the element *y*". In *A Theory of Time and Space*, we read (p. v): "The special object here aimed at has been to show that special relations may be analyzed in terms of the time relations of *before* and *after.*"

After *A Theory of Time and Space* and before *Geometry of Time and Space*. Robb (1921) published a brief monograph, *The Absolute Relations of Time and Space*, which contains, at the very end of the book, in the section "Interpretation of results," some interesting reflections, which its obvious meaning as regards Einstein's relativity (as a matter of fact, even Robb referred only to the special theory, they could be applied also to the — future — general theory) went beyond the problems he was considering (Robb 1921 77–78):

> "The important point is that our theory gives a method of setting up a coordinate system on a purely descriptive basis and introducing measurement.
>
> If we have got one coordinate system with a definite physical meaning, we can introduce any number of others.
>
> The distinction between different systems of this kind is, that while two parallel inertia lines represent the time paths of un-accelerated particles which are at rest relative to one another; two non-parallel inertia lines represent to time paths of un-accelerated particles which are in motion with uniform velocity with respect to one another.
>
> Thus our geometry makes no absolute difference between rest and uniform motion since any inertia line considered is on par with any other inertia line.
>
> It is possible of course to introduce coordinates which are various continuous functions of x, y, z, t, but it is to be noted that these have a meaning only in virtue of the meaning which we have already assigned to our original system of coordinates in terms of *before* and *after*.
>
> The change from one system f coordinates to another is equivalent to re-naming our set of elements, and may be compared to the translation from one language to another.
>
> The four numbers x, y, z, t constitute a name for an element, and if we take four functions of these variables such that x, y, z, t may be

expressed in terms of them, then these four functions may be regarded as constituting another name for the element, in a different language so to speak.

If we have a polyglot dictionary, it would be of little use to us unless we knew at least one of the languages which it contained, and a transformation from one system of coordinates to another without knowing the significance of at least one of the systems, leaves us in a similar situation to that which we might imagine an early Egyptologist would have been in who examined the Rosetta stone, but have no knowledge of Greek."

Those who know the general theory of relativity will appreciate, and be very familiar with, Robb's last statements.

1.15. The First Monographs Dedicated to Standard Relativity: L. Silberstein and E. Cunningham

1914 was, as far as the introduction of relativity in Britain is concerned, a remarkable year. Because it was then that two books dedicated to the special theory of relativity were published: Ludwik Silberstein's *The Theory of Relativity* and Ebenezer Cunningham's *The Principle of Relativity* (Silberstein 1914, Cunningham 1014). As we have seen, before 1914 discussions of Einstein's special theory appeared in some texts — the second edition (1913) of Campbell's *Modern Electrical Theory*, for instance — but such books were not dedicated exclusively to relativity, and when they were — the case of Robb —, the approach was far from the standard one. (Silberstein's *The Theory of Relativity* was result of a course of lectures dealing with Einstein's special theory of relativity that he delivered at University College London during 1912–1913.)

I shall deal briefly with Silberstein's book — which was the result of a course of lectures dealing with Einstein's special theory of relativity that he delivered at University College London during 1912–1913) — because although published in Britain and based on a course of lectures delivered at University College London, during the academic year 1912–1913, Silberstein (1872–1948) was not a British "product." A native of Warsaw, educated in Cracow, Heidelberg and Berlin, he was a lecturer,

52   *Einstein's Relativity in Great Britain*

successively, in Lemberg, Bologna and Rome, during the period 1895–1920. However, during 1912–1920 he also held in addition to his lecture-ship in Rome, positions in the Research Department of Adam Hilger, Ltd., London (a firm, founded in 1874, known for the high quality of its optical work), as well as in the University of London. Finally, in 1920 he moved to the Eastman Kodak Co. in Rochester, U.S.A., where he worked until his retirement in 1929, remaining in America until his death in 1948. Silberstein, therefore, could hardly be considered as a representative of the British scientific *milieu*. However, his book, as well as Cunningham's, was widely circulated in Britain, and this fact justifies that we mention some of its main traits.[22]

First, and foremost, let us note that Silberstein had a very fine under-standing of the real meaning of special relativity. The following quotation, taken from his book, is, by itself, an exceptionally good summary of Einstein's theory (Silberstein 1914: 87):

> "In the meantime, to the period between Lorentz's contributions and Poincaré's 1906 paper in the *Rendiconti del Circulo Matematico di Palermo* and 1905, Einstein published his paper on 'the electrodynamics of moving bodies' which has since become classical, in which, aiming at a perfect reciprocity or equivalence of the above pair of systems, S, S', and denying any claims for primacy to either, he has investigated the whole problem from the bottom. Asking himself questions of such a fundamental nature as what is to be understood by 'simultaneous' events in a pair of distant places, and dismissing altogether the idea of an aether, and in fact of any unique framework of reference, he has suc-ceeded in giving a plausible support to and at the same time a striking interpretation of, Lorentz's extended theory. Einstein's fundamental ideas on physical time and space, opening the way to modern Relativity, will occupy our attention."

---

[22] In 1924, Silberstein's book saw a second edition which included the theory of general relativity, and in 1930 he published *The Size of the Universe. Attempts at a Determination of the Curvature Radius of Spacetime* (Silberstein 1930), which dealt with relativistic cosmology (I will say more of this in Chapter 2).

The second important point is that Silberstein fully understood (as, for example, Campbell also did) that Einstein's special relativity breaks down with the notion of an ether. Thus, no reference to this concept is contained in his formulation (in p. 99) of the two basic principles of the theory. This procedure contrasts with Cunningham's or even with Eddington's 1919 presentation in his famous *Report on the Relativity Theory of Gravitation* (Eddington 1919). Commenting on the two principles, Silberstein (1914: 99) wrote:

"Fresnel claimed this property [i.e., the constancy] of light propagation only for a certain, unique system of reference, namely the aether or a system fixed in the aether, while Einstein, by accepting [the two basic axioms], postulates it for any one out of an infinity (∞^3) of systems moving uniformly with respect to one another. With regard to this property the (inertial] systems S', S'', etc., are perfectly equivalent to the system S [...] and this is the reason why the mere notion of an 'ether' breaks down."

Also relevant is mentioning that Silberstein (1914:116) realized the kinematical, universal, nature of special relativity. In this sense he stated, for instance:

"This is not to say, of course, that mechanical and all other phenomena must be 'ultimately' electromagnetic, i.e. that everything must be explained by, or reduced to, electromagnetism. The theory of relativity is not concerned at all with such reduction of one class of phenomena to another. It does not force upon us an electromagnetic view of the world any more than a mechanical view. Quite the contrary; it opens before us a wide field of possibilities of asserting that even the mass of a free electron, say a β-particle, must not be entirely electromagnetic."

That is, with Silberstein's book, British students and scientists had an excellent tool to learn special relativity. However, one thing was Silberstein's clear ideas and teachings, and a completely different one the actual expectations and understandings of the theory by some of the British audience. An anonymous review of *The Theory of Relativity*, which appeared in the *Philosophical Magazine* [29, 335–336 (1915)]

54 *Einstein's Relativity in Great Britain*

gives an idea of the terms under which British scientists took Silberstein's explanations. "The fact is — the reviewer wrote — that without the result of Michelson and Morley's experiment, there would be no justification for the theory at all. It is because it gives the most direct explanation of their null result and is at the same time not at variance with any other experimental fact, that the theory may claim serious consideration." Thinking that what he had just said were in fact Silberstein's ideas and not his own, the same commentator went on to state: "So much the reviewer felt this to be true that he would go further, and declare that it will only be when further experimental data of a crucial kind are obtained that the theory will run much chance of becoming definitely accepted as scientific knowledge."

One of the most important events in the introduction of relativity in Britain was the publication of Ebenezer Cunningham's *The Principle of Relativity*, the first book written by an Englishman and published in Britain, dedicated to the "old restricted principle," as its author called it when interviewed by John Heilbron in 19 June 1963, as part of the Oral History Project "Sources for the History of Quantum Physics", which can be read at the Niels Bohr Library of the American Institute of Physics web.[23] Fellow and lecturer of St. John's, Cambridge, when he wrote this book, Cunningham was not a major figure in physics, but there are several aspects of his career — his work in relativity being but one — which make him quite an interesting personality.[24] However, here I shall restrict myself to a discussion of Cunningham's views as contained in *The Principle of Relativity*, a book that constituted the summit of his previous original researches.[52]

Cunningham's understanding of Einstein's 1905 principle of relativity shows two, in principle, rather contradictory characteristics; namely, he

---

[23] For descriptions and some locations of the "Sources for the History of Quantum Physics," see Kuhn, Heilbron, Forman, and Allen (1967).

[24] Among these we count: "On the electromagnetic mass of a moving electron," *Philosophical Magazine 14*, 538 (1907); "On the Principle of Relativity and the electromagnetic mass of the electron: A reply to A.H. Bucherer," *Philosophical Magazine 16*, 423 (1908); "The motional effects on the Maxwell aether-stress," *Proceedings of the Royal Society of London A 83*, 109 (1909); and "The Principle of Relativity in Electrodynamics and an extension thereof," *Proceedings of the London Mathematical Society 8*, 77 (1910).

realized that the special principle of relativity is a universal one, something that not everybody understood at that epoch (in Britain as elsewhere), but at the same time he believed that such a universal principle did not entail that the ether was unnecessary. Let me quote in this connection the following passage (Cunningham 1914: 155):

> "There are two interacting processes at work in theoretical science. There is the continual endeavour to form models which [...] shall imitate or represent, more or less closely, phenomena of which the *modus operandi* is not closely realized. [...]
>
> There is on the other hand the attempt to disentangle general principles of the widest possible application, not fully descriptive of each particular set of phenomena, but common to them all, and independent of the special mechanism which is characteristic and particular. Newton's Laws of Motion must be instanced, the principle of least action, the principle of the conservation of energy. With these we may class the Principle of Relativity, that is, the general hypothesis, suggested by experience, that whatever be the nature of the aethereal medium we are unable by any conceivable experiment to obtain an estimate of the velocities of bodies relative to it."

We see from Cunningham's definition of the principle of relativity, that he had not been able to free himself from the ether concept. However, he was "modern" enough, and valued Einstein's discovery so highly, that he tried to make "ether" and "principle of relativity" compatible (we have, thus, one example of an approach in which those two concepts were not viewed as incompatible): "the critics of the Principle of Relativity — he said (Cunningham 1914: 162) — are justified in saying that it does not admit of an 'objective fixed aether,' but it cannot be said that it denies the existence of an objective aether of any kind until it is shewn that a medium cannot be conceived which renders account of electromagnetic phenomena and that at the same time has a motion which is consistent with the kinematics of the principle."

As a matter of fact, Cunningham argued — presenting also the corresponding mathematical analysis — that it is possible to conceive a *moving ether* in which the conception of stress, transfer of momentum, velocity, and flux of energy are consistently related, and which is also conformable to Einstein's kinematics, its velocity being defined by relations which have an invariant from. The magnitude of the velocity is everywhere that of light. One of the most interesting aspects of Cunningham's work concerns Minkowski's formalism, to which he dedicated a whole chapter (VIII: "Minkowski's four-dimensional calculus"), especially as a *heuristic device* to carry through the restricted relativity programme. He thus wrote (Cunningham 1914: 156):

> "The four-dimensional analysis introduced by Minkowski not only introduces a greater symmetry into the discussion of the relativity of electrodynamic phenomena. It gives us also a new point of view from which to regard mechanical quantities and enables us to go some way in finding what modifications are necessary to the usual statements of mechanical theory in order that they may be included within the scope of the principle of relativity.

Indeed, Cunningham is an example of the sympathetic reception given in Britain to Minkowski's four-dimensional formulation. In fact, after having reviewed Cunningham's main ideas concerning the basic elements of special relativity, we can go on and consider other examples of British physicists who welcomed Minkowski's formulation.

1.16. The Role Played by Minkowski's Four-Dimensional Formulation in the Introduction of Special Relativity in Britain

It is well-known that Albert Einstein had some difficulties in accepting Minkowski's four-dimensional formulation of special relativity as a positive, and interesting, contribution. (He finally became convinced of its utility and basic importance during his search for a relativistic theory of gravitation, approximately in 1912). It seems, however, that Minkowski's four-dimensional view of special relativity favoured the introduction of Einstein's theory, at least it did so in some significant cases of British physicists. We already saw, for instance, that Bateman and Cunningham held Minkowski's approach in great esteem. Moreover, they also made use of the four-dimensional framework to further some of their particular investigations (in Cunningham's case it happened thus in his studies of conformal transformations, which he even considered a possible substitute of the Lorentz transformation).[25]

We find, also, the same kind of sympathetic welcome of Minkowski's formulation in Silberstein's *The Theory of Relativity* (1914),[26] and in Eddington's *Report on the Relativity Theory of Gravitation* (1918b, 1920b), in which Minkowski's famous sentence, "Henceforth Space and Time in themselves vanish to shadows, and only a kind of union of the two preserves an independent existence," is reproduced (Eddington 1920b: 15). Indeed, Minkowski's ideas found such a congenial reception among the British, that it even appeared mixed up with theories other than Einstein's. Take, for example, the case of the Australian mathematician and mathematical physicist, but educated in Cambridge (he was the joint winner of the 1913 Adams Prize) Samuel Bruce McLaren (1876–1916), professor of Mathematics at University College, Reading. In 1913, McLaren (1913) published a paper entitled "A theory of gravity," which contains a rather complicated theory, connected, according to McLaren, with some of Bernhard Riemann's ideas, as well as with "the most revolutionary principle Physicists have accepted since Riemann's day, Einstein's principle of the relativity of time" (McLaren 1913: 636). But the point I want to emphasize here is the prominent role played by Minkowski's four-dimensional approach in McLaren's considerations: "I approach — he wrote (McLaren 1913: 639) — these problems in the first instance as a believer in the physical reality of a four dimensional. I argue that the four-dimensional geometry of Minkowski is not a mathematical fiction, but a necessary inference from ordinary experience."

1.17. Conclusions

The final acceptance of Einstein's theories came with the results of the British expedition, led by Dyson and Eddington, to measure the light bending during the solar eclipse of 29 May 1919. To this and other questions is dedicated the following chapter, but let me say now that after that

---

[25] See Cunningham (1914: 89)

[26] See especially Silberstein (1914: 129–131).

58 *Einstein's Relativity in Great Britain*

Einstein's fame was such that the extant opposition to his theories including the special theory of relativity, was considerably reduced, in Britain as elsewhere.

In some sense, the year of 1923 would be a good date to put an end to our discussion. It was then that Eddington published his masterpiece, *The Mathematical Theory of Relativity*, wherein generations of physicists and mathematicians, British or not, learned the special as well as the general theory of relativity. In fact, Eddington's treatise was but another, though eminent, example of what can be considered a general trend in the establishment of relativity in Great Britain during the 20's. Illuminating, in this sense, is the following quotation from the preface of the book *A Systematic Treatment of Einstein's Theory*, published precisely in 1923 and written by a Senior Lecturer in Physics at the University of Liverpool, J. Rice (1923: vii):

> "The demand on the part of the layman to know just what this revolution portends has been satisfied; as far as may be, by a liberal supply of popular works on Relativity. But the science undergraduate taking his normal courses in the University classroom, or reading his text-books of Physics and Mathematics, is anxious to ascertain in a more precise manner what changes this new idea is producing in the principles and content of physical science. It is primarily for him that this book has been written."

1.18. Postscript

Many of the difficulties that the Maxwellians and other physicists who reacted against special relativity were connected with the fact that Maxwell's equations were Lorentz-invariant, that is, they satisfy the requirements of Einstein's special relativity. And also the firmness with which the ether concept was installed in their minds. To say with other words, Maxwell's electrodynamics was relativistic before special relativity existed, and many did not understand that Einstein's theory was more general, stating that any physical force, not only electromagnetism, must comply with its kinematical and geometrical requirements. Hendrik

The Reception of Special Relativity in Great Britain   59

Lorentz exemplified well that misunderstanding. In his book *The Theory of Electrons*, already mentioned, the result, let us recall, of a series of lectures he delivered on Columbia University in the spring of 1906, but published only in 1909, a delay which gave him the possibility of add a few comments about developments made after 1900, he wrote (Lorentz 1952: 229–230):

> "I cannot speak here of the many highly interesting applications which Einstein has made of this principle [of relativity]. His results concerning electromagnetic and optical phenomena (leading to the same contradiction with Kaufmann's results that was pointed out in § 179) agree in the main with those which we have obtained in the preceding pages, the chief difference being that Einstein simply postulates what we have deduced, with some difficulty and not altogether satisfactorily, from the fundamental equations of the electromagnetic field. By doing this, he may certainly take credit for making us see in the negative result of experiments like those of Michelson, Rayleigh and Brace, not the fortuitous compensation of opposing effects, but the manifestation of general and fundamental principle.
>
> Yet, I think, something may also be claimed in favour of the form in which I have presented the theory, I cannot but regard the ether, which can be the seat of an electromagnetic field with its energy and its vibrations, so endowed with a certain degree of substantiality, however different it may be from ll ordinary matter. In this line of thought, it seems natural not to assume at starting that it can never make any difference whether a body moves through the ether or not, and to measure distances and lengths of time by means of rods and clocks having a fixed position relative to the other."

However, when Lorentz prepared a second edition (1915), he added the following note (Lorentz 1952: 321):

> "If I had to write the last chapter now, I should certainly have given a more prominent place to Einstein's theory of relativity by which the theory of electromagnetic phenomena in moving systems gains a simplicity that I had not been able to attain. The chief cause of my failure was my clinging to the idea that the variable t only can be considered as

the true time and that my local time t' must be regarded as no more than an auxiliary mathematical quantity. In Einstein's theory, on the contrary, t' plays the same part as t; if we want to describe phenomena in terms of $x', y'\, z', t'$ we must work with these variables exactly as we could do with x, y, z, t."

Lorentz had, finally, understood was Einstein's special theory of relativity was.

Albert Einstein, Paul Ehrenfest, Willem De Sitter, Arthur Eddington and Hendrik A. Lorentz. (Leiden, 26 September 1923). Photograph by H. Batenburg.

Chapter 2

The Reception of General Relativity Among British Physicists and Mathematicians (1915–1930)

2.1. Introduction

After considering the case of the special theory of relativity, I turn to the general theory of relativity, which was completed by Albert Einstein in November 1915 when he was working in Berlin being supported by the Prussian Academy of Sciences. Although this momentous discovery took place in the middle of World War I, when relations between many countries were interrupted, it can hardly be said that the early history of general relativity is exclusively, or even mainly, a chapter of the history of German physics. Many non-German scientists and philosophers contributed to the understanding and development of Einstein's theory. In this chapter, I discuss the contributions made to the understanding and development of the general theory of relativity by British scientists, limiting my discussion to the period between 1915 and 1930, that is, beginning with the formulation of the theory and ending with Arthur Eddington's rediscovery and acceptance of Georges Lemaitre's model of an expanding universe. At least one of the episodes of this period, the joint 1919 Royal Society and Royal Astronomical Society Eclipse expeditions to Sobral in Brazil and to Principe off the west coast of Africa, to which I will return later on, was of considerable, importance to the subsequent history of Einstein's new theory of gravitation, but it was by no means the only one; in other words,

64   *Einstein's Relativity in Great Britain*

no history of the general theory of relativity would be complete without taking into account the contributions made by British scientists.[1]

Before I proceed, let me point out that I do not include the topic, important as it is, of the reception given to Einstein's theory by British society at large. However, I would like to mention that I have never been completely satisfied with the tentative explanations of the "Einstein phenomenon" some authors have put forward. For Lewis Elton (Elton 1986), for instance, Einstein's fame arose from controversies surrounding the verification of his general theory of relativity and from the myth of its incomprehensibility, both of which originated in the Anglo-Saxon world.[23] For Marshall Missner (Missner 1985), to give another example, the American press was the instrument that made Einstein a celebrity.[3]

It might be that these explanations, among others, reveal a part of the real mechanisms behind the public reception of relativity in different national communities, however, it is difficult to understand why Einstein became so popular. Perhaps there are still some important elements missing that would allow us to solve that puzzle. What perhaps it has been not pay sufficient attention to what could be termed the "institutional history of science." To show what I mean, let me quote from a leader that appeared in the British newspaper *The Morning Post* on 27 September 1922 (it was entitled "The Scientific Front"):

> "One of the results of the war, in which the scientific brains of this country were mobilized to such good purpose, is an appreciable increase of public interest in the achievements of science, whether theoretical or practical. In prewar days, neither the man in the street nor the man at the club window could have been persuaded to read articles about the Einstein versus Newton controversy."

---

[1] The eclipse expedition, as well as other attempts, is studied in Kennefick (2019), Earman and Glymour (1980) and Crelinsten (2006).

[2] Concerning the myth of the incomprehensibility of general relativity in the Anglo-Saxon world, it is worth recalling the presumably apocryphal exchange that had Arthur Eddington as its main protagonist: "Professor Eddington — someone asked him — is it true that only three people understand general relativity?" "Oh, who is the third?" was Eddington's reply. Independently of its apocryphal or even jocular character, this anecdote sums up the prevailing opinion among a large number of individuals.

[3] For more about the cultural issues in the reception of relativity, see Glick (1987b).

If we accept the point of view expressed by this leader, then the initial reaction of the British public to Einstein and his theories, a reaction that no doubt was enhanced by the results of the 1919 eclipse expedition, ought to be looked at, at least in part, from a new perspective. Such a reaction would then be, not just the more or less straightforward response of a substantial part of the public when informed about the confirmation of a striking prediction in which the science of Britain's great hero, Isaac Newton, was confronted by that of a German Jew, Albert Einstein; rather, it would be to some extent, the fortunate coincidence of results obtained in a British-led scientific expedition with the end of a war that had definitely alerted British government and society to the values of science for the well-being and prosperity of the nation.

In this sense, it would not be very difficult, although perhaps somewhat exaggerated, to attach some kind of symbolic meaning to the 1921 publication of Richard Haldane's book, *The Reign of Relativity* (Haldane 1921); which had an extraordinary reception: It was sold out in the first week of publication, and within three months, there was a third edition. Haldane was for many years one of the most active promoters of science in Britain (see Alter 1987). Most of his efforts on behalf of science and education were made in public: within the framework of extra-parliamentary organizations, in the press, at meetings, on commissions of inquiry, and on the founding committees of new institutions (the Imperial College of Science and Technology, the National Physical Laboratory, and the Department of Scientific and Industrial Research are important examples in this sense). But they also took place behind the scenes, in negotiations and discussions with politicians and civil servants in which important decisions relating to science policy were made. I will return to Haldane in Chapter 4.

2.2. Physicists and Astronomers (1): The Introduction of General Relativity in Great Britain Before 1919

By the time — late 1915 — Einstein finally succeeded in formulating a satisfactory relativistic theory of gravitation, special relativity was already well known in Great Britain, as we saw in Chapter 1. This fact, i.e., that

66 *Einstein's Relativity in Great Britain*

Einstein's "relativity world" was by no means unknown to an important number of British scientists, should not be forgotten when one deals with the history of the reception of general relativity in Britain; it is indeed because of this fact that we can avoid the temptation of reducing the 1915–1919 period to a few events: the three papers published by the Dutch astronomer, physicist and mathematician Willem De Sitter (1872–1934) in the *Monthly Notices of the Royal Astronomical Society* (De Sitter 1916a, 1916b, 1917),[4] the role played by Arthur Eddington and Frank Dyson in the preparation and development of the 1919 Eclipse Expedition and Eddington's famous work, *Report on the Relativity Theory of Gravitation* (Eddington 1918b). There is no doubt that these events are very important, but they are only elements of a richer history.

The role played by Arthur Eddington (1882–1944), not only the introduction of relativity in Great Britain, but the development of the theory, marked much of the history of Einstein's relativity, especially general relativity, during its first decades. Extracting from A. Vibert Douglas's biography of Eddington (Douglas 1956), John Stachel (1986: 226) summarized Eddington's early interests in astronomy, and in particular in solar eclipses, as follows:[5]

"As a schoolboy of fifteen, Eddington wrote an essay on solar eclipses based on his talk to a meeting of the Brymelyn Literary Society a few months earlier. Five years later, during his second year at Cambridge (1903), he read a paper on 'the velocity of gravitation' to the Trinity Mathematical Society, and in 1906 he read a paper on gravitation to the $\nabla^2 V$ Club.[6]

---

[4] A few aspects related to De Sitter's three papers in the *Monthly Notices* were discussed by A.D. Crommelin (1931) in the address he delivered on the occasion of the award of the Gold Medal of the Royal Astronomical Society to De Sitter. For a biography of De Sitter, see Guichelaar (2018).

[5] Besides Douglas's (1956) biography of Eddington, valuable information is contained also in Stanley (2007) and Chandrasekhar (1983).

[6] Eddington took the Tripos examinations dafter only two years of study. When the results were declared, it came out that Eddington was Senior Wrangler; never before had a second-year student won such distinction.

> Although he studied physics at Cambridge, in 1906 he accepted a position as Chief Assistant at the Royal Observatory in Greenwich. There he served his apprenticeship in the astronomical profession between 1906 and 1913. [...]
>
> It was in 1912 that he obtained his first chance to participate in a solar eclipse expedition."

Indeed, well before De Sitter appeared on the horizon in 1916, Eddington had participated in 1912 in a Brazilian solar eclipse expedition that included in its research program a test of Einstein's 1911 prediction of the apparent position of a star in the gravitational field of the sun. The hypothesis that Eddington knew — presumably of Einstein's 1911 work from Charles Dillon Perrine, an American astronomer who had become the director of the Argentine National Observatory in 1909 — is supported by the fact that he referred to it in print in an article published in February 1915 (Eddington 1915).

As a matter of fact, Eddington's reference to Einstein's 1911 theory came too late, even for British scientists, since in the January 1915 issue of *Philosophical Magazine* Adriaan D. Fokker, as mentioned in Chapter 1 one of Lorentz's students, had already published a paper explaining Einstein and Marcel Grossmann's 1913 theory of gravitation (Fokker 1915). Fokker, who had collaborated with Einstein for a few months during 1913–1914 in a problem dealing with Gunnar Nordström's 1912 theory of gravitation (Einstein and Fokker 1914), submitted his paper in July 1914 when he was visiting William H. Bragg in Leeds (it was Bragg who communicated Fokker's paper to *Philosophical Magazine*). The Dutch physicist, who at that time was very much interested in Einstein's ideas on gravity, spent some time in England in 1914, making the acquaintance of Paul Dirac, Alfred A. Robb, and Ernest Rutherford, as well as of Bragg. In fact, it seems quite likely that he could have discussed Einstein's ideas with some of them.[7]

---

[7] See the interview conducted by John L. Heilbron with Fokker, as part of the "Archive for History of Quantum Physics. Sources for History of Quantum Physics" project (Kuhn, Heilbron, Forman, and Alen 1967).

68   *Einstein's Relativity in Great Britain*

Almost as soon as he began to receive information on Einstein's final theory, through the good services of De Sitter, Eddington helped to organize a discussion on gravitation, which was held at the annual meeting of the British Association for the Advancement of Science.[8] In 1916 (September 5–9), the British Association, which was overshadowed by the war, (no meetings were held in 1917 and 1918) met at Newcastle-on-Tyne. Unfortunately, the report of the meeting mentions (*Report of the Eighty-Sixth Meeting of the British Association for the Advancement of Science* 1917: 364) only that the discussion took place and that it was opened by Ebenezer Cunningham; we also know that Eddington discussed Einstein's new theory of gravitation.[9] Thus, as early as the summer of 1916, many British scientists had the opportunity of hearing about general relativity.[17]

Soon afterward, and well before the famous 1919 eclipse, the Lindemanns, Adolph (father), and Frederick Alexander (son, and 1st Viscount Cherwell), published a paper (submitted on 4 December 1916, and published in the December issue of *Monthly Notices of the Royal Astronomical Society*) dealing with daylight photography of stars as means of testing the equivalence postulate of the new theory of relativity (Lindemann and Lindemann 1916). They indicated methods for testing by photography Einstein's hypothesis "of the refraction of light by a gravitational field," as they called it; they also discussed the most favourable conditions for photographing stars in the daytime, describing experiments that showed that stars may be photographed in daytime using red filters and red sensitive plates. Moreover, they described an instrument by means

---

[8] This was before De Sitter sent Eddington his first paper for the *Monthly Notices*. De Sitter's paper was completed during August 1916, and in July, Eddington was writing to the Dutch astronomer that "I feel sure you will allow me to make use of the papers you send."

[9] See Eve (1939: 254), and the account of the BAAS meeting published in *Nature* 98: 120 (1916). Eddington's talk at Newcastle-On-Tyne presumable coincides with his first published paper devoted to the general theory of relativity (Eddington 1916). Among those present at Newcastle-On-Tyne were F.W. Dyson, E. Rutherford, J.C. McLennan, H.R. Hasse, W.H. Hicks, H.H. Tumer, and A.N. Whitehead, who presided over Section A (Mathematics and Physics).

The Reception of General Relativity 69

of which the transparency of the air at different times and places may be compared quantitatively. As their conclusion, the Lindemanns suggested that "experiments in daylight photography of stars be undertaken by some observatory possessing a suitable instrument, and enjoying a fine climate, with a view to testing Einstein's theory." It was, as far as I know, the only contribution made by the Lindemanns to the development of general relativity, but it shows that the early reception of Einstein's theory in Great Britain included more protagonists than the ones usually mentioned.

During 1917, things began to speed up. Eddington prepared a council note that the Council of the Royal Astronomical Society included in its report to the 97th Annual General Meeting, held on 9 February 1917.[10] Eddington's note, entitled "Einstein's theory of gravitation" (Eddington 1917a), left no doubt as to how he viewed general relativity:

> "A council Note in 1910 begins with the sentence, 'Celestial mechanics, which has hitherto been based on the Newtonian laws of motion, is profoundly affected by the discoveries which have been made in recent years regarding measurements of space, time, and force' [*Monthly Notices*. 70: 363). These words were of the character of a prediction, for the new ideas had at that time scarcely reached the principal force with which astronomy is concerned-gravitation_ But, remarkable progress has been made, especially in the last two years, and now there emerges a complete and self-consistent new system of mechanics, which can be applied without ambiguity to the problems of gravitational astronomy."

In the same volume of *Monthly Notices*, Frank Dyson (1868–1939), the Astronomer Royal, wrote about the opportunity that the coming eclipse of 29 May 1919, offered for verifying Einstein's theory of gravitation (Dyson 1917). It was on 21 September 1910, when Dyson, then Scotland Astronomer Royal, received a letter from W. Graham Greene, of

---

[10]The council notes, or, more properly, the "Reports on the Progress of Astronomy," were commissioned by the Council of the Royal Astronomical Society. It was only from 1938 that the authors' names were spelled out, although the owners of the initials used until then were easily identifiable.

70   *Einstein's Relativity in Great Britain*

the Admiralty, announcing his appointment as Royal Astronomer at Greenwich (Wilson 11951: 148–149):

"Sir,
I am commanded by the Lords Commissioners of the Admiralty to acquaint you that the King has been pleased to approve of you being appointed as Astronomer Royal to succeed Sir William Christie, K. C. B., D. Sc., F. R. S., and that you will required to undertake the duties of that office from the 1st proximo, which is the date Sir William Christie desired to resign the appointment.
Your salary will be £1,000 a year, with the official residence and My Lords are informed that you have stated that you will prepared to retire on reaching the age of 60."

In the biography of her father, Margaret Wilson (1951: 191) stated that "At far back as March, 1917, when the war situation still looked desperate and the chance of sending eclipse expeditions abroad hardly worth considering, Dyson had drawn attention to the fact that the eclipse of 29 May 1919, might well be of unique importance, as it would provide an opportunity of testing Einstein's Theory of Relativity, which, if let pass, would not occur again for nearly twenty years. Throughout 1918 he went steadily ahead with all the necessary preparations. As Chairman of the Joint Permanent Eclipse Committee (a committee of the Royal society and the Royal Astronomical Society), he was the moving spirit of the whole undertaking."

In his 1917 paper, and besides speaking of the possibilities this eclipse would offer, Dyson mentioned that he had already tried to use four photographs taken with the astrographic equatorial al the 1905 eclipse by C.R. Davidson, an astronomer of the Royal Observatory in Greenwich who had been one of Eddington's assistants in the 1912 Brazil eclipse expedition and who would also be on the 1919 expedition (he went to Sobral with Crommelin). The results did not support Einstein's theory, but, as Dyson pointed out, the observations had not been designed with the purpose of testing a gravitational theory. Dyson's conclusion was that the observations show, however, that astrographic object-glasses are very suitable for this important observation.

The Reception of General Relativity   71

Another event that helped introduce general relativity to Great Britain during 1917 and 1918 was a discussion that took place in the pages *of Philosophical Magazine*, with Oliver Lodge and Arthur Eddington as the main protagonists and a small contribution from G.W. Walker (Walker 1918), but I will discuss it in Chapter 3.

1918 was an important year as far as the introduction of Einstein's relativity in Great Britain because it was then when it was published the first edition of Eddington's famous article, *Report on the Relativity Theory of Gravitation* (Eddington 1918b). Prepared at the request of the Physical Society of London, the report was the first complete account of general relativity to appear in the English language. It is worth to quote what Eddington said in the Preface:

> "Whether the theory ultimately proves to be correct or not, it claims attention as being one of the most beautiful examples of the power of general mathematical reasoning. The nearest parallel to it is found in the applications of the second law of thermodynamics, in which remarkable conclusions are deduced from a single principle without any inquire into the mechanisms of the phenomena; similarly, if the principle of equivalence is accepted, it is possible to stride over the difficulties due to ignorance of the nature of gravitation and arrive directly at physical results. Einstein's theory has been successful in explaining the celebrated astronomical discordance of the motion of the perihelion of Mercury [Einstein value 43]" without introducing any arbitrary constant; there is no trace of forced agreement about this prediction. It further leads to interesting conclusions with regard to the defection of light by a gravitational field, and the displacement of spectral lines of the sun, which may be tested by experiment."

The success of Eddington's report was great, especially after the results of the 1919 Eclipse became known, and another edition, modified with those results, appeared in 1929.

The scene was, therefore, ready for the great year, the year of the eclipse, a year that marked the irrefutable success in Great Britain of Einstein's theory. To some extent, it was, as we have seen, both the end of

72   *Einstein's Relativity in Great Britain*

and the sequel to a three-year period that saw a significant number of discussions, analyses, and publications.[11]

To end this section, 1918, the year previous to the eclipse, was the first year that the contributions of Ludwik Silberstein to the analysis of general relativity were published (Silberstein 1918a, 1918b). I will return to Silberstein later; let me just say that he was most probably at that time, with the possible exceptions of Eddington and De Sitter, the physicist publishing in Britain with the broadest and deepest knowledge of general relativity, although it is true that at that time he did not fully accept Einstein's theory.

2.3. 1919: The Year of the Eclipse and the Astronomers

Much has been written about the famous joint meeting of the Royal Astronomical Society and the Royal Society, held on Thursday, November 6, 1919, in the rooms of the Royal Society at Burlington House. A large

---

[11] This view of the 1915–1918 history of general relativity in Great Britain, although not essentially compatible with the one put forward by Earman and Glymour (1980), offers new perspectives and insights concerning the early reception of Einstein's 1915 theory in Britain. Compare, in this sense, what I have said before with the following statements made by Earman and Glymour (1980: 50): "Yet there is no evidence that before 1919 many British physicists had warmed to general relativity. What little literature appeared on the theory was generally critical, or concerned with presenting alternatives. Before 1919 there did not appear in British scientific journals a single article, other than De Sitter's, that applied or extended the new theory. British physicists had to take cognizance of the theory, primarily because Eddington and De Sitter had made them aware that Einstein had succeeded in explaining the long-standing anomaly in the motion of Mercury, but prior to the eclipse expedition they were not disposed to accept the new relativistic account of gravitation or even to trouble much to understand it." To further my point, I will recall the case of William Wilson, of King's College, London, who, as can be seen in bibliographies such as the ones prepared by Combridge (1965) and Lecat (1924) or in Whittaker's encyclopedic history (Whittaker 1953), contributed to the development of general relativity during the 1920s and 1930s. By 1918, Wilson was already discussing specific points of Einstein's theory at the Physical Society of London (see Wilson 1919); in this paper, received at the Physical Society on 21 October 1918, Wilson showed that the equations of motion in a gravitational field can be put in a Hamiltonian form.

audience, consisting mainly of the fellows of both societies, assembled with the president of the Royal Society, J.J. Thomson, to hear the Astronomer Royal (Dyson) and Eddington's accounts of the eclipse expedition. The meeting was in fact a "meeting for discussion,"[12] and although the conclusion put forward by Dyson and Eddington, namely, that the results obtained were in full agreement with the prediction for the bending of light in Einstein's general relativity, was generally accepted, one speaker, Silberstein, dissented, and E.A. Lindemann reported on a pilot experiment that aimed at detecting the deflection of infrared radiation by the Sun outside eclipses.

It is famous the description that Alfred North Whitehead (), who attended the meeting, made of the presentation of the eclipse results in his book *Science and the Modern World*:

> "The whole atmosphere of tense interest was exactly that of the Greek drama: we were the chorus commenting on the decree of destiny as disclosed in the development of a supreme incident. There was dramatic quality in the very staging: — the traditional ceremonial, and in the background the picture of Newton to remind us that the greatest of scientific generalisations was now, after two centuries, to receive its first modification. Nor was the personal wanting: a great adventure in thought had at length come safe to shore."

Indeed, the announcement of the results obtained in the eclipse expedition caused a tremendous sensation in England. Eddington, from them on the real champion of relativity in England. Headlines, such as the one appearing in the London *Times* of November 7, "Revolution in Science. New Theory of the Universe/ Newtonian Ideas Overthrown," were not infrequent. But, it was not only the daily press or the scientific journals that paid attention to the news; magazines, such as *The Athenaeum* and *The Nineteenth Century and After*, also included in their pages commentaries dedicated to the new gravitational theory.[13]

---

[12] See, for instance, the note published in *The Athenaeum*, 21 November 1919 (p. 1229), as well as Tayler, ed. (1987: 26).

[13] See, for instance, *The Athenaeum*, "Einstein's Theory of Gravitation," 14 November 1919, p. 1189, and "A Matter of Evidence," 21 November 1919, pp. 1128–1129.

74   *Einstein's Relativity in Great Britain*

On December 11, Eddington wrote to Einstein informing him the reception given to his theory in England (CPAE 2009: 158–159):

> "Our results were announced on Nov. 6 and you probably know that since then all England has been talking about your theory. It has made a tremendous sensation and although the popular interest will die down, there is no mistaking the genuine enthusiasm in scientific circles and perhaps more particularly in this university. It is the best possible thing that could have happened for scientific relations between England and Germany. I do not anticipate rapid progress towards official reunion, but there is a big advance towards a more reasonable frame of mind among scientific men, and that is even more important than the renewal of formal associations. [...] I have been kept very busy lecturing and writing on your theory. My Report on Relativity is sold out and is being reprinted. That shows the zeal for knowledge on the subject because it is not an easy book to tackle. I had a huge audience at the Cambridge Philosophical Society a few days ago, and hundreds were turned away unable to get near the room. Although it seems unfair that Dr. Freundlich, who was first in the field, should not have the satisfaction of accomplishing the experimental test of your theory, one feels that things have turned out very fortunately in giving this object lesson of the solidarity of <British> German and British science even in time of war. I, likewise, am unable to write except in my own language. Yours sincerely, A. S. Eddington."

The public reception of Einstein's theories did not diminished with time. A god example in this sense is another latter Eddington wrote to Einstein, this time on 11 February 1929 (CPAE 2021: 598):

> "My dear Einstein,
>
> I was glad to receive reprints of your papers ['Zur einheitlichen Feldtheorie,' *Preussisch Akademie der Wissenschaften (Berlin). Physikkisch-mathematische Klasse- Sitzungsberichte*, 2–7 (1929)]. I very much hope that it is a sign that your health is better than when you wrote me last.
>
> I think I have understood, and I written a short article for 'Nature' ["Einstein's field theory," *Nature 123*, 280–281 (1929)] which I hope will be of assistance to English students who are trying to understand your paper.

The Reception of General Relativity 75

> You may be amused to hear that one of our great Department Stores (Selfridges) has pasted up in its windows your paper (the six pages up side by side) so that passers by can read it all through. Large crowds gather round to read it."

And he added, in a different tone: "But I am afraid *you have not converted me* at present. We have discussed on former occasions whether anything further could be expected in getting a natural connection between electricity and gravitation by field-theory alone. I don't think it will come out this way; but if the working out of your field-law to a second approximation yields anything interesting I would reconsider this opinion,"

In a paper published in the magazine *The Nineteenth Century and After*, even Oliver Lodge (1919), a truly "classical physicist" — I will return to this in the next chapter —, accepted without any reservation the measurements taken by the British expedition, pointing out also that "Before Einstein's prediction nothing of the kind had been seen, nothing of the kind had been looked for, nor, so far as is known, has such an amount of deflection been suspected." However, he could not refrain from warning of the dangers of accepting Einstein's theory. "The present writer holds it dangerous — he wrote — to base such far-reaching consequences [i.e., Einstein's ideas of space and time as expressed in his gravitational theory], even if anything like them can legitimately be drawn — which is doubtful — on a predicted effect which may after all be accounted for and expressed in simpler fashion. Our admiration for the brilliant way in which the fact was arrived at must not make us too enthusiastically ready to assimilate the whole complicated theory out of which it arose. For, independent of the above speculation, it is indubitable that the mathematical theory of relativity-very different from what philosophers have hit her to meant by the term-is almost inevitably complicated." Lodge also took this opportunity to mention his "Electrical Theory of Matter," to which I have already referred.

The full account of the eclipse expedition results was published by Dyson, Eddington, and Cottingham (1920) in the *Philosophical Transactions of the Royal Society of London*. It might be worth pointing out at this stage that a preliminary discussion of part of the eclipse results (the photographs taken at Príncipe) was presented by Eddington and Edwin Turner Cottingham (1920) at the 87th meeting of the British

76 *Einstein's Relativity in Great Britain*

Association for the Advancement of Science, held at Bournemouth on 9–13 September 1919. The meeting and discussion of November 6 was not the only scientific discussion organized in England soon after the official announcement of the eclipse results: The Council of the Royal Astronomical Society decided that its own meeting of December 12 should also be devoted to a discussion of general relativity[14]; the Royal Society also organized a discussion that took place on 5 February 1920; the next month (March 26), it was the turn of the Physical Society of London.[15]

In a sense, the least interesting of these meetings was the one organized by the Physical Society of London. Indeed, it was but a sort of dialogue between Eddington, who gave a general presentation of the theory, and those in the audience, who wanted to ask him some questions. Eventually, 10 persons did so. Joseph Larmor sent a communication that was read by the secretary of the society; in it, Larmor criticized Einstein's theory, of which he said "How far it is from being a determinate theory of the universe will appear to anyone who dips into the writings of its developers." At the same time, Larmor took the opportunity to point out the compatibility of general relativity with his beloved principle of least action. Indeed, Larmor, of which I will say more in the next chapter, was quite active in the discussions that followed the 1919 announcement; being, however, a truly 19th-century scientist, he could not fully accept Einstein's theory. A representative sentence in this sense is the following one, which appears in the paper Larmor wrote in collaboration with W.J. Johnston for the 1919 meeting at the British Association for the Advancement of Science (Johnston and Larmor 1920): "If physical

---

[14] Usually, the ordinary meetings of the Royal Astronomical Society were devoted to the reading of papers. However, in view of the widespread interest in the theory of relativity caused by the publication of the results of the eclipse observations, the Council thought they would meet the wishes of the fellows by giving the whole time of the meeting of December 12 to the consideration of Einstein's theory. Regarding this, see Dreyer and Turner, eds. (1923: 238).

[15] See "Discussion on the Theory of Relativity," *Monthly Notices of the Royal Astronomical Society 80*, 96–119 (1919); "Discussion 'On the Theory of Relativity," *Proceedings of the Physical Society of London 32*, 245–251 (1920).

science is to evolve on the basis of relations of permanent matter and its motions, time must continue to be thrown on to the material observing system in the form of slight modification of its structure." This paper will appear again in Chapter 3.

As for the other two meetings, several speakers, Eddington, J. Jeans, Dyson, Silberstein, and E.A. Lindemann, spoke at both, with Lodge, H. Jeffreys, and Larmor (the latter with a contributed paper) participating only in the Royal Astronomical Society meeting, and R.H. Fowler, E. Cunningham, and H.F. Newall in the one organized by the Royal Society.

There were some common traits between the Royal Astronomical Society and Royal Society meetings; Eddington and Jeans gave introductory talks (at the Royal Astronomical Society meeting, Jeans confined practically all of his comments to the special theory of relativity; perhaps his knowledge of general relativity was still rather superficial), and the Astronomer Royal made brief comments dealing with observational matters. Of all the speakers participating in both events, the most critical was Silberstein, discounting Larmor and Lodge. He questioned the British astronomers' conclusions, observing "displacements from the rectangular components quoted by the Astronomer Royal at the joint meeting of November 6." Although he was "aware of the difficulty of these measurements and of the circumstance that the direction is a too severe test for them," he could not avoid concluding that if "we were not prejudiced by Einstein's theory, we should not have said that the figures strongly indicated a radial law of displacement." He argued that even if the "displacements were ideally radial, we may have to account for them in some other, non-Einsteinian way."

One of those non-Einsteinian ways was considered by Silberstein in a paper that appeared in the February 1920 issue of *Philosophical Magazine* (Silberstein 1920). In his article, he stated that it was premature to interpret the bending of light effect found by the British astronomers as a verification of Einstein's theory, "not merely in view of the small outstanding discrepancies, but chiefly in view of the failure of detecting the spectrum shift predicted by the theory, with which the whole theory stands or falls." To Silberstein, the eclipse result proved that there was an alteration,

78   *Einstein's Relativity in Great Britain*

a change of light velocity all around the sun, and it was at this stage that he turned to the Stokes–Planck ether:

> "The condensation claimed by Planck's modification of Stokes's theory, for the Sun as well as for the Earth and for all other material bodies, is no longer devoid of influence on observable phenomena. It suddenly acquires physical life, so to speak. In other words, the discovery made in Brazil naturally suggests the idea that the observed deflection *is due to the condensation of the ether* around the sun, and that although one has been an implacable enemy of any ether at all for the last 15 years, one does not hesitate to point out this possibility, a last glimpse of hope, perhaps, for the vanished medium."

Silberstein's revival of the Stokes–Planck's ether gave Lodge (Lodge 1920) the opportunity to call attention to an old paper (1907) of his, but nothing else came out of it.

Coming back to the Royal Astronomical Society meeting, we note that Silberstein's most important point there was his insistence on the lack of any observation confirming Einstein's theoretical prediction of a gravitational displacement toward the red of the solar spectral lines "Einstein's whole theory — he remarked — stands or falls with this effect."

Although agreeing on the importance of the red-shift prediction, Lindemann did not think that the absence of such a shift had been proved; to support his contention, he quoted the conflicting values found by a number of researchers (C.E. St. John, L. Grebe and A. Bachem, and K. Schwarzschild).[16] However, Harold Jeffreys' reading of the observed spectral shifts was somewhat different from Lindemann's; comparing Einstein's theoretical value with those measured by John Evershed and Charles St. John, he concluded that "while they dont show any uniform agreement with the theory [i.e., general relativity], they are at least sufficiently in agreement with it to suggest that it cannot be condemned on the spectroscopic evidence alone." At the Royal Society meeting, Alfred Fowler took another view, pointing out that different "interpretations may

---

[16] On this regard, as well as about what follows, see Crelinsten (2006).

The question of the spectral shift, the third of general relativity tests, became a topic frequently discussed in England, as elsewhere, during the 1920s. The problem was that the testing of this prediction was a very difficult matter, not so much because the small displacement of 0.008 A was difficult to measure, but because it was masked by larger displacements of an unknown amount. Scientists such as Duffield (1919–1920) studied the effect, but it was mostly because of the efforts of John Evershed (1864–1956), an F.R.S. British astronomer — he was the first to observe radial motions in sunspot — attached to Kodaikanal Observatory, a solar observatory owned and operated by the Indian Institute of Astrophysics, and of Charles Edward St. John (1857–1935), of the Mt. Wilson staff — probably, his most important work was his revision of the Rowland's table of solar wavelengths undertaken during the 1920s in collaboration with other Mt. Wilson colleagues — that it became accepted, although not without a certain amount of doubt, that the observations agreed with the theoretical prediction derived from general relativity.

Briefly, the story is as follows: from his observations at Kodaikanal in India, Evershed had convinced himself by 1921 that the solar spectrum yielded evidence for the gravitational displacement of lines to the red in accordance with the requirements of Einstein's theory (Evershed 1921). But St. John, whose first observations were made in 1917, with his magnificent equipment at Mt. Wilson was still dubious in 1922, and outlined the difficulties and conflicting evidence in his address at the Centenary Meeting of the Royal Astronomical Society on 30 May 1922, with Eddington in the chair as president (Douglas 1956: 44–45). However, one year later, St. John (1922–1923) thought he had confirmed the displacement to the red within 14% of the Einstein value. The chief cause of his change of attitude was his recognition of the very small pressure in the reversing layer of the solar atmosphere so that it was no longer necessary to confine the study to lines that were insensitive to pressure shifts. In view of these results, the report of the Council of the Royal Astronomical

---

[17] The observations referred to by Fowler was made between 1914 and 1919 by Evershed, Schwarzschild, St. John, and Grebe and Bachem.

80 *Einstein's Relativity in Great Britain*

Society at its 104th General Meeting included the statement[18]: "The measures of the solar spectral lines no longer offer any difficulty to the existence of the Einstein shift, even to those who hesitate about adopting this explanation of the observed facts." Nevertheless, it was an overly optimistic assessment; thus, during his address as the new president of the Royal Astronomical Society in 1926, Jeans (1925–1926) pointed out that "Even today the problem can hardly be said to be solved with absolute finality so far as solar light is concerned, although probably Evershed, St. John, and others have done all that it is possible to do." The new president thought that 'the case was different, however, with the light coming from one of the faintest of stars, Sirius, because of the fact that Sirius has a smaller and far fainter star as a companion, describing an orbit around it. The corresponding measurements were undertaken at Mt. Wilson, and Jeans remarked that "the observed shift was found to agree almost exactly with that predicted by theory. This experiment not only established the validity of Einstein's theory of the gravitational displacement of spectral lines, but also showed its usefulness as an instrument to be utilized by the practical astronomer in his everyday work." Again, not everybody would have subscribed to Jeans's words, although, certainly, the available observations were regarded as weighting the evidence for Einstein's theory against that of the sceptics.[19]

From what I have said, it seems obvious that the reaction of Einstein's 1915 was coped by the astronomers, but what about the British physicists, when such distinction can be made?

In Chapter 1, I mention an episode involving Emest Rutherford — his 1910 comment that "no Anglo-Saxon can understand relativity; they have too much sense." However, by 1919 Rutherford's feelings as regarding relativity, at least as far as general relativity was concerned, seem to have changed; thus, the December 25 issue of *Nature* contained a note entitled "Radioactivity and Gravitation," co-authored by Rutherford and Arthur H. Compton (Rutherford and Compton 1919), in which they reported the

---

[18] "The Astronomical Tests of Einstein's Law," *Monthly Notices of the Royal Astronomical Society* 84: 292–295 (1924). The author of this report answered to the initials A.C.D.C., most probably A.C.D. Crommelin, who was the Society's secretary during the period 1917–1923.

[19] The problem continued to be investigated: see, for instance, Evershed (1928, 1931) as well as Eisenstaedt (2003: Chapter 9) for how the test evolved.

The Reception of General Relativity 81

results of some experiments performed in the Cavendish Laboratory to test whether the rate of decay of radioactive substances is affected by subjecting them to the high centrifugal acceleration at the rim of a spinning disc.[20] The motivation of this investigation, which Arthur Schuster had suggested to Rutherford before the outbreak of the war, was to test whether the rate of transformation of a radioactive substance was influenced by the intensity of gravitation, 38 a problem that became particularly interesting in view of the fact that, as they put it, "according to Einstein's theory, a gravitational acceleration is in no sense different from a centrifugal acceleration."[21] Rutherford and Compton's contribution was, however, purely experimental; they did not make any effort to see whether or not their experimental results agreed with the predictions of general relativity, most probably because they were unable to deal with the complexity of the new theory.

Very different was Paul Dirac's reception, which he recalled in his speech at the symposium held in Jerusalem in 1979 to commemorate the centennial of Einstein's birth (Dirac 1982: 79–80):

"The time I am speaking of is the end of the First World War. That had been long and terrible. [...] Then the end of this war came, rather suddenly and unexpectedly, in November 1918. There was immediately an intense feeling of relaxation. It t was something dreadful that was now finished. People wanted to get away from thinking about the awful war that had passed. They wanted something new. And that is when relativity burst upon us. [...]

At that time I was sixteen years old and a student of engineering at Bristol University. [...]

I was caught up in the excitement of relativity along with my fellow students. We were studying engineering, and all our work was based on Newton. We had absolute faith in Newton, and now we learned that Newton was wrong in some mysterious way. This was a very puzzling situation. Our professors were not able to help us, because no one really had the precise information needed to explain things properly, except for one man, Arthur Eddington."

---

[20] Rutherford and Compton's note dealt, as they acknowledged, with the same sort of problems that the chemist Frederick George Donnan was studying at the time (see Donnan 1919). Indeed, their results were "not in disaccord with the relation deduced by Prof. Donnan."

[21] Of course, this was in fact a test of the principle of equivalence.

82 *Einstein's Relativity in Great Britain*

On such account, it is not surprising that Dirac's second paper concerned relativity dynamics, in which he showed that what Eddington had called kinematic and dynamics velocities were identical (Dirac 1924b), or that in 1975 he published a schematic brief treatise about general relativity (Dirac 1975).[22]

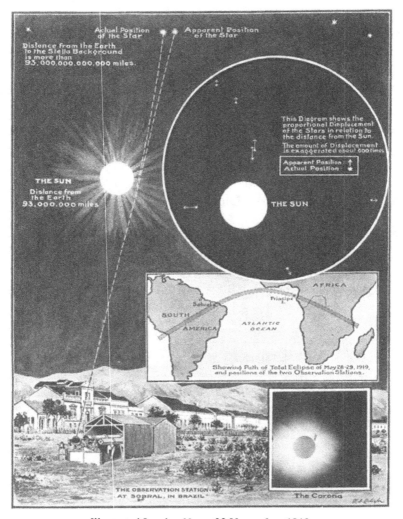

Illustrated London News, **22 November 1919**

---

[22] His first paper (Dirac 1924a) dealt with dissociation under a temperature gradient.

2.4. Arthur Eddington's *Space, Time & Gravitation* and *The Mathematical Theory of Relativity*

General relativity in Great Britain during the period 1920–1930 was dominated by the towering figure of Arthur Eddington. Although it is not my intention to analyse his works and ideas here, I can hardly proceed without mentioning some of his contributions.

Eddington's activities in support of general relativity can be divided into several groups. As far as books are concerned, after publishing *Report on the Relativity Theory of Gravitation*, he saw that there was a need for a less mathematical account, which would explain as many as possible of the underlying ideas and results of relativity theory using a minimum of mathematical symbols. Thus, in 1920, he published his third book: *Space, Time & Gravitation* (Eddington 1920c)[23]; "One of the literary gems of English scientific literature," James Rice (1923) called Eddington's book (soon a French edition was published: Eddington 1921a). In the "Preface, Eddington explained that:

> "It is my aim to give an account of this work [Einstein's] without introducing anything very technical in the way of mathematics, physics or philosophy. The new view of space and time, so opposed to our habits of thought, must in any case demand unusual mental exercise. The results appear strange; and the incongruity is not without an humorous side. For the first nine chapters the task is one of interpreting a clear-cut theory, accepted in all its essentials by a large and growing school of physicists — although perhaps not everyone would accept the author's views of its meaning. Chapter x ['Towards infinity'] and xi ['Electricity and Gravitation'] deal with recent advances, with regard to which opinion is more fluid. As for the last chapter, containing the author's speculations on the meaning of nature, since it touches on the rudiments of a philosophical system, it is perhaps too sanguine to hope that it can ever be other than controversial."

---

[23] The same year, 1920, appeared the English translation — by Robert W. Lawson, of the Physics Laboratory, Sheffield University — of Einstein's *Über die spezialle und die allgemeine Relativitätstheorie* (1917): *Relativity, the Special & the General Theory. A Popular Exposition* (Methuen, London).

84 *Einstein's Relativity in Great Britain*

The French edition of *Space, Time & Gravitation* included a mathematical supplement that was in fact a draft of Eddington's masterpiece, The *Mathematical Theory of Relativity* (Eddington 1923). In the "Preface", Eddington (1923: v) acknowledged this connection:

> "A first draft of this book was published in 1921 as a mathematical supplement to the French Edition of *Space, Time and Gravitation*. During the ensuing eighteen months I have pursued my intention of developing it into a more systematic and comprehensive treatise on the mathematical theory of Relativity. The matter have been rewritten, the sequence of the arguments rearranged in many places, and numerous additions made throughout; so that the work is now expanded to three times its former size. It is hoped that, as now enlarged, it may meet the needs of those who wish to enter fully into these problems of reconstruction of theoretical physics."

For many years, this treatise was the standard textbook from which many physicists and mathematicians learned the special and general theories of relativity.[24] Of this book, Paul Dirac (1982: 82) said:

> "We really had no chance to understand relativity properly until 1923, when Eddington published his book, *The Mathematical Theory of Relativity*, which contained all the information needed for a proper understanding of the basis of the theory. This mathematical information was interspersed with a lot of philosophy. Eddington had his own philosophical views, which, I believe, were somewhat different from Einstein, but developed from them.
>
> But there it was, and it was possible for people who had a knowledge of the calculus, people such as engineering students, to check the work and study it in detail. The going was pretty tough. It was a harder kind of mathematics than we had been used to in our engineering training, but still it was possible to master the theory. That was how I got to know about relativity in an accurate way."

---

[24] A second edition was published in 1924 and reprinted in 1930, 1937, 1952, 1957, 1960, 1963 and 1965. Also, a German translation appeared in 1925; it contained an appendix written by Einstein.

Besides the standard general theory of relativity, Eddington's book contained chapters on the special theory of relativity, tensor calculus, cosmological models, electricity and gravitation, and what Eddington called "world geometry", which included Weyl's theory and other attempts to generalise general relativity, and task to which Eddington joined (see, for instance, Eddington 1921b).[25] On the cosmological realm, it is worth remembering that Eddington (1923: 161) showed that in de Sitter cosmological model particles "will not remain at rest unless [they are] at the origin, but will be repelled from the origin with an acceleration increasing with the distance, a property that, Eddington clearly realized, could perhaps be used to solve "one of the most perplexing problems of cosmogony, [that of] the great speed of the spiral nebula."

Thus, *The Mathematical Theory of Relativity* was important not only because of its "didactic dimension" (i.e., because it could be regarded as a textbook in which the basic elements of Einstein's gravitational theory were put forward), but also because it contained quite a number of what were then advanced topics, most of them advanced topics he had previously published in specialized journals, but their inclusion in *The Mathematical Theory of Relativity* increased their popularity among a larger audience, thus aiding in their development.

During years, Eddington gave a course of lectures on general relativity each Easter term in Cambridge; according to Douglas (1956: 51), "a few senior undergraduates and research students and a few of the younger fellows took advantage of these lectures. About eight attended the course in 1922." However, his role as university professor and as public lecturer were very different as regards influence. "Relatively — Douglas (1956: 107) explained — few students in any one year attended Eddington's lectures in the University classrooms; far more heard the occasional special addresses which he was always willing to give to any group or society; thousands of people across four continents had the experience of hearing one or more of his public lectures. These last were invariably well prepared, written out in full, well ordered, highly informative, enriched with literary quality and with his inimitable humour."

---

[25] I dealt with some aspects of Eddington's attempts of generalizing Weyl's approach in Sánchez-Ron (1987 b: 176–180).

86 *Einstein's Relativity in Great Britain*

Turning now toward Eddington's own original contributions to the development of general relativity, we can say that during the period considered here his most important works were those dealing with gravitational waves and with the generalization of Weyl's unified theory of electromagnetic and gravitational fields (Eddington 1921b, 1922). Indeed, both topics received attention from some British scientists: for example, the papers that O.R. Baldwin and G.B. Jeffery of University College London, published dealing with the relativity theory of plane and divergent waves (Baldwin 1929, Balwin and Jeffery 1926), and articles by McVittie, Whittaker, and Wirtinger related to unified field theories (see Combridge 1965).

It will be obvious to many readers that I have not mentioned Eddington's ideas which he exposed in his books *Relativity Theory of Protons and Electrons* (Eddington 1936), his posthumous *Fundamental Theory* (Eddington 1953), his *The Combination of Relativity Theory and Quantum Theory* (Eddington 1943), a product of lectures at the Dublin Institute for Advanced Studies, or his more philosophically-oriented, *The Nature of the Physical World* (Eddington 1928), or *The Philosophy of Physical Science* (Eddington 1939).[26] The reasons why I have not taken into account this part of Eddington's work — and Milne's theory, of which I will say something later on — were well explained by Helge Kragh (2011: 108–109) in one of his books, *Higher Speculations. Gran Theories and Failed Revolutions in Physics and Cosmology*: "Seen over a longer time span, the theories of Eddington and Milne were just failures, two more examples of attempted revolutions in physics that ended up in the graveyard of theories of everything. By 1950 the two theories of cosmophysics were largely forgotten and current interest in them is essentially restricted to historians and philosophers of science. While Hermann Bondi found some of the elements in Milne's theory valuable, he dismissed Eddington's way of thinking. A deductive theory that aims at showing that an observational known fact is a necessary consequence of the process of human thought is immediately suspected of fitting the ends

---

[26] Clive Kilmister (1966) has reviewed Eddington's works among them his efforts to combine relativity with quantum theory.

to the end. This was undoubtedly a common attitude among physicists, not only after Eddington's death but also during his lifetime."

2.5. More on the Reception of General Relativity

However, the number of research papers published in this process of assimilation and development of general relativity in Great Britain is rather small, especially when one considers what the situation was like in other fields of physics. Thus, if we review the papers published in British scientific journals during the period 1920–1930,[27] we find that the number of articles dealing with general relativity published by British physicists, astronomers, and mathematicians — it really makes no difference to consider also non-British scientists — was almost insignificant compared with the number of papers with quantum physics as their subject, something which was not strange, as the problems of quantum physics at the time absorbed the energies of an important part of theoretical and experimental physicists (in England, the school, or influence, of Rutherford was dominant) A possible way of illustrating this is by looking at the reports of the British Association for the Advancement of Science meetings. If we take into account the meetings held between 1921 (Edinburgh) and 1930 (Bristol), we find that, according to the corresponding reports, only two papers dealing with general relativity were read, namely, Silberstein's "Determination of the curvature radius of space-time" (Toronto 1924; p. 373 of the *Report* published in 1925), and S. Brodetsky's "The equations of the gravitational field in two and in three dimensions of space-time" (Leeds 1927; p. 315 of the *Report* published in 1927), the latter being read within the Department of

---

[27] As far as journals are concerned, my discussion throughout this paper is based on the analysis of the following journals: *Proceedings of the Royal Society of London* (series A), *Philosophical Magazine, Monthly Notices of the Royal Astronomical Society, Philosophical Transactions of the Royal Society of London, Proceedings of the Royal Society of Edinburgh, The Journal of the London Mathematical Society, Proceedings of the Cambridge Philosophical Society, The Quarterly Journal of Pure and Applied Mathematics* (after Vol. 50, its name was *The Quarterly Journal of Mathematics*), and *The Memoires and Proceedings of the Manchester Literary and Philosophical Society.* I have also been helped by the bibliographies prepared by Lecat (1924) and Combridge (1965).

88 *Einstein's Relativity in Great Britain*

Mathematics of Section A (Mathematical and Physical Sciences). To these two papers, we ought to add the already mentioned in Chapter 1 lecture delivered by the mathematician Edmund Whittaker (1927), as president of Section A of the 1927 Leeds meeting. Whittaker's lecture, and the one delivered in 1921 by Owen Richardson (also in Section A) were the only addresses, including the presidential ones, delivered during those years that mentioned Einstein's theory (in Whittaker's case, that theory was the theme of the lecture; in Richardson's case, Einstein's theory was only briefly mentioned).[28] When one compares this with the number of addresses, papers, or discussions delivered, read, or held at the British Association forum that dealt with quantum physics issues (for example, with the analysis of crystal structure by X-rays, or with the analysis of line spectra), one sees how poorly represented general relativity was or, what amounts to the same thing, where the real strengths and potentialities of British physics laid.

If we now look at the topics treated in the general relativity articles published by British scientists, setting aside those papers considered in the last section dealing with the classical experimental tests of the theory, we find that most dealt with electromagnetic phenomena (the electromagnetic field of the electron, in particular) in a gravitational field a theme with which many classical scientists could feel relatively at home.[29] Following in the list of often-treated topics was what we could term mathematical aspects of general relativity, but as the authors of these works were usually mathematicians, I will delay my discussion of them until the next section. There were also papers dealing with other topics (with the problem of motion, for example), but the numbers involved were rather small.

---

[28] Indeed, even when he was mentioning general relativity, Richardson could not avoid referring to quantum physics in very favourable terms. Thus, he said (Richardson 1922): "Relativity is the revolutionary movement in physics which has caught the public eye, perhaps because it deals with familiar conceptions in a manner which for the most part is found pleasantly incomprehensible. But it is only one of a number of revolutionary changes of comparable magnitude. Among these we have to place the advent of the quantum, the significance of which I hope we shall thoroughly discuss early next week."

[29] See, for example, items numbers 277, 655, 667, 731, 734, 1000, 1671, and 1677 of Combridge (1965), corresponding to papers of Conway, Cunningham, Hargreaves, Jeffery, Meksyn, E.T. Whtttaker, and J.M. Whittaker.

As far as journals are concerned, and although *Philosophical Magazine*, the *Proceedings of the Royal Society of London*, and even the *Proceedings of the Cambridge Philosophical Society* included in their pages an important number of general relativity papers, it was the *Monthly Notices of the Royal Astronomical Society* that played the most important role in the introduction and development of Einstein's gravitational theory in Great Britain. *Nature* was also important; the 17 February 1921, issue, for example, was totally dedicated to relativity; it included articles by Einstein, Cunningham, Dyson, Crommelin, St. John, Mathews, Jeans, Lorentz, Lodge, Weyle, Eddington, Campbell, Wrinch and Jeffreys, and Carr. And we should not forget *The Observatory*, the monthly review that aimed presenting in a popular form the progress of astronomy.[30] However, I have in general excluded *Nature* and *The Observatory* from my considerations because they were more scientific magazines than scientific journals, and I am only considering the latter ones here.

Besides the articles published in these journals, there were also the council notes dealing with general relativity matters that were included in the Royal Astronomical Society journal. We have discussed some of them, but now I will mention another one that coincides with the end of the period considered in this chapter. In many ways, Eddington's discovery or, better, rediscovery and acceptance, of George Lemaître's 1927 model of an expanding universe constituted the end of a period in the history of general relativity in Great Britain; the universe was then to be seen, especially in Eddington's case, in a new light, a light that opened up fresh problems that a new generation of young physicists (such as George McVittie — a student of Eddington, to whom is dedicated Chapter 5 of the present book —, William McCrea, and Subrahmanyan Chandrasekhar), who had grown up with relativity as a well-established theory, could tackle. Eddington's new convictions were evident in his 1930 paper, "On the instability of Einstein's spherical world" (Eddington 1930a),[31] but a more public pronouncement was the council note he prepared early the

---

[30] For a list of the articles dealing with general relativity that appeared in *The Observatory*, see the entry "Relativity" in *The Observatory. Generalities*, Vols. 1–72 *(1877–1955)*, pp. 108–109.

[31] See also Eddington (1930b).

90 *Einstein's Relativity in Great Britain*

following year, under the attractive title "The expansion of the Universe" (Eddington 1931). Finally, it is worth pointing out that in the index of the 80th volume (1919–1920) of the *Monthly Notices*, there was a section dedicated to relativity; this fact clearly shows the status attained by Einstein's theory in the minds of the authorities of the Royal Astronomical Society (it is interesting to recall that Eddington was the president of this society during the period from 1921 to 1923).

Turning to monographs dedicated to general relativity, we see that Eddington's books were not alone. Indeed, the first half of the 1920s saw the publication of a large number of books, usually textbooks or general introductions that had Einstein's relativity as their leitmotiv. Leaving aside the English translations of foreign books (a partial list can be found in the "Books and Pamphlets" of the "Bibliography of Relativity" published in the 17 February 1921, *Nature* issue dedicated to relativity (pp. 811–813). as well as pamphlets and books of a too general nature or of a philosophical character (Chapter 4 is dedicated to the reception of relativity among British philosophers), we can mention a relatively small number of books, some of them already mentioned in Chapter 1, Cunningham's *Relativity, the Electron Theory, and Gravitation* (Cunningham 1921), Rice's *Relativity: A Systematic Treatment of Einstein's Theory* (Rice 1923), Herbert Dingle's *Relativity and Gravitation* (Dingle 1923), Percy Nunn's *Relativity and Gravitation: An Elementary Treatise upon Einstein's Theory* (Nunn 1923),[32] Norman Campbell's *Relativity* (a supplementary chapter to his 1907–1913 *Modern Electrical Theory* (Campbell 1923), and Silberstein's *The Theory of Relativity* (Silberstein 1914), *The Theory of General Relativity and Gravitation* (1922) and *The Size of the Universe* (Silberstein 1930) though the last two books were published when he was installed in the United States, that did not mean that they were not read in Great Britain (actually *The Size f the Universe* was published by Oxford University Press).

Adding what was said of Silberstein in Chapter 1, it should be pointed out that he never lost his independent way of looking at general relativity

---

[32] Nunn, of whom I will say more on Chapter 4, was professor of Education at the University of London. I have included him here, in spite of his not being really a scientist, because his book shows a considerable technical knowledge of relativity.

problems, but we should not confuse such an independence with a rejection of the theory, at least when referring to him around 1922. The "Preface" he included in *The Theory of General Relativity and Gravitation* (Silberstein 1922) is expressive in this sense: "Some of my readers will miss, perhaps, in this volume the enthusiastic tone which usually permeates the books and pamphlets that have been written on the subject (with a notable exception of Einstein's own writings). Yet the author is the last man to be blind to the admirable boldness and the severe architectonic beauty of Einstein's theory. But it has seemed that beauties of such a kind are rather enhanced than obscured by the adoption of a sober tone and an apparently cold form of presentation."[33] Another expression of his independent, and often unsuccessful, approaches can be found in his controversial conception of cosmology, which he summed up in *The Size of the Universe* (Silberstein 1930), written before Lemaître's model of the universe became known. When Eddington (1930b) reviewed Silberstein's book, he called attention to Lemaître's "very substantial advance" which "renders obsolete the Contrast between Einstein's and De Sitter's cosmologies" on which Silberstein based his work (see in this regard Kragh 1987).

---

[33] In that "Preface" he also explained the origin of the book: "At the Conference on Recent Advances in Physics held in the Physics Laboratory of the University of Toronto from 5 to 26 January 1921, a course on Einstein's Relativity and Gravitation Theory, consisting of fifteen lectures and two colloquia, was delivered by the author. The first six of these lectures were devoted to what is known as Special Relativity, and the remaining ones to Einstein's General Relativity and Gravitation Theory and to relativistic Electromagnetism. In view of the time limitations only the essentials of these theories were dealt with, due attention, however, being given to the critically conceptual side of the subject. The University was kind enough to undertake the publication of that part of the course, which dealt with general relativistic questions, on the express understanding that my prospective readers should be assumed to be already familiar with the special theory of relativity. In this connection it was suggested by Prof. McLennan that those unacquainted with the older theory should be referred to my book of 1914 (*The Theory of Relativity*, Macmillan, London) and that it would therefore be desirable to make the present volume, as much as possible, uniform in exposition and style with that work. With such requirements in view this little book was shaped, only a few pages at the beginning having been used in recalling the essentials of the special relativity theory."

92   *Einstein's Relativity in Great Britain*

2.6. General Relativity and British Mathematicians

The mathematician's response to the general theory of relativity is a particularly interesting topic, in as much as Einstein's theory requires mathematical tools that are not needed usually in other fields of physics. Indeed, it is a well-known fact that the contributions of many people who had been educated as mathematicians proved to be very important for the development of the general theory. As a matter of fact, many physicists, and philosophers by the way, seemed to perceive Einstein's theory as belonging in some sense to the domain of mathematics. Thus, Oliver Lodge's remark, made during the already mentioned December 1919 "Discussion on the Theory of Relativity" held at the Royal Astronomical Society, could be considered as representative of the feeling of at least some of his colleagues[34]:

> "The surprising thing is that this theory has arrived at verifiable results; it is marvelous, and it represents very brilliant mathematical work. It has been done by using an out-of-the-way calculus developed by pure mathematicians-a ponderous kind of tool which only a few people can use. I do not pretend to be able to use it; I only with difficulty follow it; but the principle must coincide with some kind of reality, for by writing down equations on that principle you can get results. The theory is not dynamical. There is no apparent aim at real truth. It is regarded as a convenient mode of expression."

Of course, we know that Lodge was not a sympathetic commentator (more on Chapter 3), yet, he was not unfair completely to Einstein's gravitational theory. It is true that at that time few people mastered the absolute calculus needed to understand the theory, and its experimental dimension, although impressive, was rather slim. Therefore, there were reasons that could justify the physicists' discomfort with the theory, especially in the case of physicists not too worried with fundamental principles.

With mathematicians, the reaction was very different. General relativity, a scientific theory that was even discussed in newspapers, gave them new opportunities, and some mathematicians were able to use their abilities

---

[34] See note 13; Lodge's remark appears on p. 107.

to contribute to the development of a theory about which scientists and the general public were speaking. General relativity offered them a "feeling of life" so to speak, which they could hardly find in other domains of mathematics or mathematical physics. Many physicists had to radically change, when confronted with Einstein's theory, their education and (conservative) ideas of what physics was really about; however, with mathematicians, it was almost the opposite. People such as Andrew Russell Forsyth, Arthur Cayley's successor in the chair of Pure Mathematics at the University of Cambridge (later on, in 1910, he moved to Imperial College, London), arry Edward James Curzon, an obscure mathematician who was lecturer in mathematics at Goldsmiths' College, London, and Edward T. Copson, lecturer in mathematics at the University of Edinburgh and later Regius Professor of Mathematics at the University of St. Andrews, could participate (or thought they did) in the spectacular new quest to know the structure of the universe. Forsyth, who in 1912 had published a book on the differential geometry of curves and surfaces (Forsyth 1912), was the author of two papers Forsyth (1920, 1921) in which he applied his considerable knowledge of differential equations (see Forsyth 1890–1906) to solve some of the equations appearing in general relativity. Curzon (1924a, b) published two papers in the *Proceedings of the London Mathematical Society*, a society of which he was a member, which contained the so, called "Curzon Exact Solution" of Einstein's field equations, a solution still often discussed in some of the modern general relativity texts. As far as I know, these two important papers of Curzon (the first is really only an abstract) constitute his only contributions to the development of general relativity. And Copson (1928) dealt with "the electrostatics in a gravitational field" in a paper published in the *Proceedings of the Royal Society of London*.

A particularly illustrative case in this sense is that of John Edward Campbell (1862–1924). A mathematician attached for most of his life to Hertford College, Oxford, Campbell worked on miscellaneous topics. By 1897, he was writing about matters connected with Lie's transformation groups, which were his main concern for several years, culminating in his book *Introductory Treatise on Lie's Theory of Finite Continuous Transformation Groups* (Campbell 1903), a book that somewhat dampened John Theodore Merz's criticism of British mathematics, as expressed

94 *Einstein's Relativity in Great Britain*

in his classic *A History of European Scientific Thought in the Nineteenth Century*.[35] The geometrical applications of Lie's contact transformations gradually led Campbell to the differential geometry of surfaces, and papers on Backlund's transformation, the application of quaternions to the deformation of a surface, and cyclic congruencies were written by him during the 10 years after the publication of his book. Then, his research came to an end. He only recovered his interest in mathematics after the war; it was tensor calculus and the applications of differential geometry to Einstein's theory that appealed to him, although he never had shown any interest either in mathematical physics or in experimental science.[36] In 1922 and 1923, he published two articles dealing with purely mathematical aspects of general relativity (Campbell 1922, 1923). Moreover, elected president of the London Mathematical Society in 1920, he chose as the subject of his presidential address (read on 11 November 1920) "Einstein's theory of gravitation as a hypothesis of differential geometry" (Campbell 1920). He also wrote a book that was posthumously published by Clarendon (Oxford University) Press under the title *A Course of Differential Geometry*

---

[35] There, Merz (1904: 690–601) wrote: "The 'Theory of Groups' has now grown into a very extensive doctrine which according to the late Prof. Marius Sophus Lie [...] is destined to occupy a leading and central position in the mathematical science of the future. [...] But though it is an undoubted fact that the largest systematic works on the subject emanate from the great Norwegian mathematician, and that his ideas have won gradual recognition [...] the epoch-making tract which pushed the novel conception into the foreground was Prof. F. Klein's 'Erlangen Programme' (1872). [...] From that date onward, the different kind of groups have been defined and systematically studied, notably by Klein and Líe and their pupils. In this country [Great Britain], although many of the relevant ideas were contained in the writings notably of Cayley and Sylvester, the systematic treatment of the subject was little attended to before the publication (1887) of Prof. Burnside's 'Theory of Groups of Finite Order,' and latterly of his 'article on the whole theory of groups in the 29th volume of the 'Ency. Brit.' It has been remarked by those who have studied most profoundly the development of the two great branches of mathematical tactics — viz. 'The Theory of Invariants' and the 'Theory of Groups' — that the progress of science would have been more rapid if the English School had taken more notice of the general comprehensive treatment of Lie."

[36] For more information about Campbell, see the obituary note (signed by H.H.) that appeared in the *Proceedings of the London Mathematical Society 23*, lxx–lxxi (1924).

(Campbell 1926); unfortunately, as his son (J.M.H. Campbell) wrote in the preface, he died before completing a chapter, or appendix, that was to deal with the connection between the rest of the book and Einstein's theory.

General relativity served another obvious purpose for British mathematicians: through it some of them became acquainted with new areas of mathematics, such as parallelism and teleparallelism (or distant parallelism), a topic to which Edmund Whittaker (1930) dedicated his 1929 presidential address to the London Mathematical Society.[37] In Great Britain, during the period 1915–1930, there were no mathematicians of the category of Weyl, Levi-Civita, Cartan, Schouten, of Veblen, who could promote the development of differential geometry in a significant way; at this time, British mathematics was stronger in other fields, such as analysis, mathematical logic, and applied mathematics, its principal names including Hardy, Littlewood, Whitehead, Russell, Whittaker, Forsyth, Watson, Lamb, Love, and Pearson.[38] There is no doubt that the accomplishments of British mathematicians in differential geometry were not terribly impressive (especially when compared with the achievements of mathematicians in other countries), but the rather small number of papers dealing with different aspects (differential invariants, tensor calculus, curves in Riemannian spaces, and so on) of this area of mathematics, papers that were published at least in part because of general relativity, helped in the diversification of British mathematics. In this way, general relativity contributed to the development of mathematics in Great Britain, a phenomenon not exclusive to the British case, but felt more intensively in Great Britain than in other countries.

As far as individual British mathematicians are concerned, it was probably Edmund Taylor Whittaker (1873–1956), professor of Mathematics at the University of Edinburgh from 1912, who achieved the greatest distinction for his contributions to the development of general relativity

---

[37] Here is Whittaker's definition of 'teleparallelism': "When the vector obtained at the terminus is the same whatever be the path by which the journey from the initial point has been made, we say that there is *teleparallelism*" (Whittaker 1930).

[38] As I said, Eddington contributed to some of these mathematical topics, but I am not including him here because he was not a mathematician.

96 *Einstein's Relativity in Great Britain*

(about topics like Hilbert's world function, electrical phenomena in gravitational fields or unified theories). Nevertheless, it could be argued that John Synge's contributions to relativity were more numerous and important than Whittaker's. John Lighton Synge (1897–1995) was an Irishman, who studied in Dublin, first at St. Andrew's College, and then, since 1915, at Trinity College, where he became lecturer, until in 1920 when he accepted a position as assistant professor of Mathematics at the university of Toronto, in Canada, where he attended lectures by Ludwik Silberstein on the theory of relativity. In 1925, he returned to Trinity College Dublin where he became university professor of Natural Philosophy. However, five years later he went back to Toronto as professor of Applied Mathematics and head of the Department of Applied Mathematics. And his itinerant career did not end there: he spent time and held positions at Princeton University (1939–1941), Brown University (1941, Ohio State University (1943–1946), and Carnegie Institute of Technology in Pittsburgh, before he returned to Ireland in 1948, as Senior Professor in the School of Theoretical Physics at the Dublin Institute for Advanced Studies, that had been set up in 1940. His influence in the development of general relativity in Great Britain would be more notable later, both through his papers and through people such as Felix A.E. Pirani, who spent some time with him in Dublin, after what he went to King's College, London.[39] I will say more of Synge at the end of Chapter 4.

Whittaker's case is remarkable not only because of his general relativity papers, all of them concerned in some way with electromagnetism, but also because at Edinburgh he did what his colleagues in the physics or, better, natural philosophy department (Charles G. Barkla and Charles Galton Darwin were the professors there) did not: he lectured on Einstein's theory of relativity to senior year undergraduates, and post-graduates. In the late 1920s, for example, he lectured on the unification of gravitation and electromagnetism. In the academic year 1927–1928, a graduate student who followed this course was George McVittie, to whom will be dedicated Chapter 5.

---

[39] The list of Synge's publications is included in O'Raifeartaigh, ed. (1972). There, one can see the papers dealing with general relativity, or with mathematical problems related to Einstein's theory, that Synge published in British journals during the 1920s.

However, there were not many British mathematicians or physicists who contributed so actively to the development and assimilation of general relativity as did Whittaker. John Synge (1958: 39, 46) summed up beautifully Whittaker's constributions to relativity[40]:

"Ten papers deal with the theory of relativity, and all are concerned in some way with electromagnetism. Those who know Whittaker through his *Modern Analysis* and *Analytical Dynamics* will recognise in these papers the same mastery over complicated situations which enabled him to disdain the support of notational refinements, that same elegance, brevity and persuasive charm which make difficult arguments seem easier than they really are. The new element which emerges is the strongly geometrical approach; but he remains true to the Lagrange tradition and draws no diagrams of space-time, although these must surely have been before his mind's eye and would have helped his readers. [...]

I see Whittaker as a consummate artist in the formal symbolism of mathematics, his passion being to make the symbols dance to his tune and to take their place in persuasive arguments of great economy, inessential details being suppressed. But this passion alone did not suffice for he yearned on the one hand towards that precision of mathematical thought which was lacking among the British mathematicians of his age (a situation which he did much to remedy), and on the other hand he yearned for contact with the realities of the physical universe — realities which, in their broader aspects, he would treat with philosophical depth, but which he was not adapted to uncover for himself. He could narrate in a most lucid way the history of electromagnetism, both mathematically and physically, but the passion for narration and the passion for creation are very different. He had, it seems, no consuming passion for physical reality, no more than Lagrange or Hamilton had; he was more akin to them intellectually, than to Newton or Maxwell or Kelvin or Rayleigh."

These last sentences could be applied well during decades to many scientists who worked in relativity as if it were a mere chapter of mathematics.

---

[40] For Whittaker's contributions to relativity, see also Temple (1956).

98 *Einstein's Relativity in Great Britain*

2.7. Milne's Kinematical Relativity

In all I have said before, a name is absent, that of Edward Arthur Milne (1896–1950), the astrophysicist and mathematician who from January 1929 until his death in September 1950 held the Rouse Ball chair of Mathematics at Oxford University. Besides his contributions to astrophysics, he was well known in his time due to what he called "kinematical relativity," which he presented in several works, especially in two books: *Relativity, Gravitation and World-Structure* (Milne 1935) and *Kinematical Relativity; a Sequel to Relativity, Gravitation and World-Structure* (Milne 1943).[41] However, I will not say much of Milne's theory mainly because it fell outside the time period I am considering: its earliest formulation was in 1932 (Milne 1932), though it is true that it aroused much attention since its appearance, especially during the second half of the 1930s and early 1940s.[42] Kinematical relativity implied the abandonment of some of the fundamental assumptions of general relativity, in particular, the principle of covariance; basically, it was based on two postulates, which Urani and Gale (1993: 398) stated as follows: (1) acceptance of the constancy of the speed of light, which was equivalent to accepting special relativity and Lorentz invariance, and that cosmology should be described in a flat Lorentzian space-time; and (2) that two observers in relative motion should have identical views of the universe (both local and inferred views). In the "Foreword" that Roger Penrose (2013: x–xi) added to the biography that Meg Weston Smith (2013), Milne's daughter, wrote of her father, he said:

> "Milne's own distinctive approach to cosmology, which he referred to as *kinematic relativity*, paid scant attention to Einstein's motivation for

---

[41] See also, Milne's idiosyncratic book, *Modern Cosmology and the Christian Idea of God*, based on the Edward Cadbury Lectures he delivered in the University of Birmingham in 1959.(Milne 1952). Also worthwhile is the text of the presidential address that Milne delivered at the meeting of the London Mathematical Society of 15 December 1939, who he dedicated to explain the kinematical relativity (Milne 1940).

[42] A good review of Milne's theory is included in North (1990: Chapter 8). See also Gale and Urani (1999), which contains a discussion of some of the controversies kinematical relativity aroused, and the influence that kinematical relativity exerted on Bondi's philosophy in the steady state theory.

The Reception of General Relativity 99

regarding gravitation as a feature of space-time curvature, and was concerned basically with the space-time geometrical aspects of a spatially uniform expanding Universe. Although Milne's theory ignored he revolutionary insights that Einstein had introduced concerning gravitation, it was by no means simply a retreat to the older Newtonian ideas of space and time, providing instead a novel and far-reaching outlook concerning the basis of space-time-geometry. One aspect of this was to concentrate on *time* rather than space — the notion of separation that is ascertain by *clocks*, rather than by rulers — as being the basic measure of space-time geometry. Light signals provided the link that enabled space measures to be derived through the use of *radar*-type determinations. This very insightful idea of basing space- time geometry on time measurements rather than spatial ones (which had, in essence, also been previously put forward by the leading Irish relativity theorist John Lighton Synge in a brief paper published in *Nature* in 1921, and often promoted by him later) was, at the time, met with much skepticism, and even derision."

In Chapter 5, Milne's kinematical relativity will appear again in connection with George McVittie.

2.8. Epilogue

On the whole, the great majority of the scientists who in one way or another contributed to the reception and development of general relativity in Great Britain during the period 1915–1930, did so from time to time; Einstein's theory constituted a small part of their professional interests, something, of course, that should be understood in the light of the fact that general relativity was a completely new theory, created by the genius of essentially one person, Albert Einstein. Consequently, the process of its institutionalization as a distinctive branch of physics, with its own experts (scientists working almost exclusively on its development), took some time, especially in a scientific community with the characteristics of the British one. During 1919 and 1920, the discussions could easily be intense because of the novelty and experimental dimension that the eclipse results brought to the theory, but afterward, the situation changed radically; new experimental effects were difficult, if not impossible, to find or deal with, and the development of general relativity relied mainly on its more theoretical and mathematical aspects. Few British physicists and mathematicians had the abilities, technical skills, opportunities for discussion with other colleagues, and time to work systematically and exclusively on the development of general relativity, especially in view of the many opportunities offered at the time by the successful atomic theory. By 1930, however, the situation began to change, with a new generation of scientists who had grown up with Einstein's theory as an already established theory.

Chapter 3

Larmor, Lodge and General Relativity: The "Old Guard" Reaction to Einstein's Relativity

When Thomas Kuhn interviewed John Cockcroft on 2 May 1963, as part of the project "Sources for the History of Quantum Physics," they began to talk about the Cambridge the famous British physicist had known when he was a young student, early in the 1920s. Referring to Arthur S. Eddington, Cockcroft had no doubt that "he was a big figure," and that "naturally, we went to [his] lectures." Edward A. Milne, also, was well considered, especiallyw because of his work on the internal constitution of the stars; and in the field of quantum physics, there was the great Ernest Rutherford, and others, like Ralph H. Fowler, his son-in-law. But what about the older generation? J.J. Thomson was still around, and the youngsters "used to go to [his] lectures," not really to learn new physics, but "for the physics of the early twentieth century [...] electrons and all this kind of things; discharge in gases, and isotopes. It was very nice to have him there and to be able to go and listen to the great man." It seems, therefore, that the old 'J.J.' had some audience among Cambridge students; indeed, the instruction he gave did not conflict at all with the direction atomic physics had taken, in as much as he did not express himself against any new ideas, at least according to Cockcroft's recollections.

Another "great man" of the University at the time was Joseph Larmor, whom we have met already in Chapter 1 and also, briefly in Chapter 2.

102 *Einstein's Relativity in Great Britain*

Cockcroft, who seems to have been a diligent student, attended Larmor's lectures, but there he would find only two or three more students. "He represented," Cockcroft recalled, "the physics of the 1900s. He was a grand old man and it was worth going to hear him. Sometimes he would come out with some new ideas, such as the transmission of electromagnetic waves by the ionosphere and he would come out with those in his lectures before they became generally known. It was interesting to go to but he really did not represent the new age in theoretical physics at all."[1] Indeed, he seemed a man whose heart was in the nineteenth century, with the names of Faraday, Maxwell, Kelvin, Hamilton, Stokes ever on his lips, as though he mentally consulted their judgement on all the modem problems that arose. He would often say that all true scientific progress ceased about 1900 or even earlier for his own *fin de siècle* effusion was only dubiously judged by the looser standards of these times; but that was as far as he would go, except when he forgot his pose. There was, of course, a great deal of exaggeration in this pose; but he adopted it so systematically that perhaps he himself could scarcely distinguish it from his natural opinions.[2] But which were "his natural opinions"? In this chapter, I will try to discover his opinions concerning Einstein's general theory of

---

[1] Larmor himself seems to have recognized such feelings. Thus, on 9 September 1924, he wrote to Oliver Lodge that "I have acquired the idea that I am suddenly old and passé, through being bombarded by the officials (older than me) with formal returns to fill up as to the conditions on which I must retire on a pension now, at 70, and what they can offer." However, he was not prepared to leave Cambridge: "I really would retire if I could only retain a fellowship." Then he added: "They will make Cambridge into a Government bureaucracy yet. The younger generation takes no interest, Jets it all slide." Unless otherwise stated, all the letters mentioned in this chapter are deposited in the "Lodge Collection", The Library, University College London, MS Add 89. I am grateful to the librarian for allowing me to have access to this depository.

[2] Other commentators have put forward similar opinions. Thus, Buchwald (1981b: 374) wrote: "In his later years he was reknowned as a staunch conservative in his scientific views, often delighting in the role of the older scientist defending the Olympian past against the immodest pretensions of the present." Similarly, Woodruff (1981: 39) stated that "unlike Lorentz, Larmor did not participate to a larger extent as a guide to the newer generation of physicists developing quantum theory and relativity. In general, he maintained a conservative, critical attitude toward the new ideas, particularly examining the possible limitations of the relativity theories."

relativity, by using mainly — though not only — his correspondence with Olíver Lodge, whose points of view will also be studied. At the same time, further comments will be made about the points of view of Arthur Eddington, James Jeans and Edmund Whittaker, who corresponded with Lodge.

Larmor's history as regards Einstein's relativity is not one of successes, but of frustrations and final defeat. It is the story of how an able and intelligent physicist, brought up during the nineteenth century, reacted when confronted with such a radical new theory as general relativity.

3.1. Larmor and Lodge and the Introduction of Relativity in Great Britain before 1919

We had seen already that the early reception of relativity in Britain was a complicated process and that a turning point was the 1919 eclipse expedition led by Frank Dyson and Arthur Eddington. However, it is important to point out that it was not only the new generation of physicists and mathematicians who learnt about relativity quite early on but also the 'old guard,' people like Larmor and Lodge, who were also quite well informed. Thus, on 20 October 1916, Larrnor wrote to Lodge:

> "Einstein is so dissatisfied with an aether and motion relative to it that he makes a gravitational field a modification of surrounding space from the Euclidian form which a body carries about in rigid connexion with it! And our friend Lorentz is fascinated by it, rather as a malhematical exercise I think than a philosophical theory."

That is, by 1916, Larmor knew of the existence of general relativity. To which extent he understood that theory is something we shall consider later on.

Of course, to some extent, it is not surprising that both Larmor and Lodge were interested in anything new Einstein had to say, in as much as their own work had dealt in the past — we already saw this — with questions that Einstein's previous theory, special relativity, had illuminated in a form different from what they had expected. This is the case, for

104   *Einstein's Relativity in Great Britain*

instance, with Larmor's already mentioned *Aether and Matter*, as well as with Lodge's 1892 experiments in Liverpool, aiming at determining whether the ether could be set in motion.[3]

Contrary to what one could have perhaps expected, it seems that Larmor and Lodge did respect Einstein, although they clearly did not share his points of view (more about this later). Writing on January 21, 1915, to a friend, J. Arthur Hill, Lodge, the less knowledgeable and more hostile in the end of the two, stated (Hill 1932: 61–62):

> "In speaking of Einstein and the other upholders of the doctrine, I ought to add that they are brilliant mathematicians and learned people, and not to be sniffed at lightly — that is what has caused their doctrine to catch hold. They also appear to be sustained by experiment to this extent that all those who attempted to attack the question of Matter through Ether- and they have been numerous-have had a negative result. The Principle of Relativity assumes that a negative result they always will and must have, no matter how far they are pushed."

3.2. Problems with the Mathematics of General Relativity

Contrary to Larmor's case, Lodge found general relativity rather difficult to understand from the very beginning, especially because of his limited grasp of the mathematics involved. And this contrast between two antagonists to relativity deserves our attention because it allows us to understand better the nature of Larmor's opposition to relativity, rooted in physical and philosophical considerations, and not on mathematical difficulties. A decade after the eclipse expedition, on 27 May 1929, Lodge confessed to Edmund Whittaker his mathematical shortcomings[4]:

> "I was grateful to you for sending me your lecture on "What Energy Is?[5] But I am rather horrified to find that I cannot follow, i.e., understand, it all.

---

[3] Lodge (1892). For a historical analysis of Lodge's experiments, see also Hunt (1986).

[4] For the relation of Lodge with mathematics, see also Stanley (2020).

[5] Whittaker (1929). This article is based on a lecture Whittaker gave at the Edinburgh University Physical Society on 16 March 1929.

It rather shocks me that the tensors have to be introduced in connexion with so fundamental a thing as energy. I don't even know what a tensor is. I know that a vector is a scalar with direction as well as magnitude. One has got used to vectors. I suppose a tensor is a vector with something added. But what? Is it a twist, or what Robert Ball called a wrench? At my age I am not going to learn the tensor calculus whatever happens. And I am rather surprised that conservation of energy has to be muddled up with conservation of momentum in order to make a complete statement. [...]

It is the concluding part of your lecture which makes me rather hopeless about understanding the modem view. I feel as if there must sooner or later be some simpler way of specifying fundamental things. Matrices and tensors are not the kind of mathematical weapons which I can imagine posterity using with any satisfaction, even though they are interim necessities. But I look to you to go beyond interim necessities and not to leave us in the quagmire prepared by Dirac and others.[6] It is very little comfort to admit that matter is a form of energy, if energy turns out to be nothing concrete but only a mathematical abstraction."

Lodge's statements represent a minor historical problem. Why was he so ignorant in, at least, those branches of mathematics? Was it because he went to University College London (UCL), instead of going to Cambridge and taking the Mathematical Tripos? However, in London, he had received instruction in "Riemann surfaces and many other parts of higher mathematics" from Olaus Magnus Henrici, a German mathematician who became professor of Pure Mathematics of UCL, and there was also W.K. Clifford, Professor of Applied Mathematics (Lodge 1932: Chapter 6). Obviously, sooner or later, he simply forgot those teachings if he ever learnt them. Why he forgot — if that was the case — is a matter of guesswork, but one possibility is because the London method of examination did not put the same emphasis on the resolution of problems as did the Mathematical Tripos (the aspirants to wranglers had to solve hundreds of problems to sustain any hope of obtaining a good position in the Tripos). But, of course, this is only a possibility, sustained also,

---

[6] It is worth recalling that one year before, Paul Dirac (1928) had used matrices in his relativistic equation of the electron.

106 *Einstein's Relativity in Great Britain*

it is true, by the fact that people like Larmor or Eddington, who could understand the mathematics of general relativity, were former wranglers.[7]

Whatever problems Larmor had with general relativity, its mathematics was not one of them. Apart from mathematical ignorance, there was also a rather British dislike for "merely mathematical" theories, based on the belief that physics was not the same as mathematics, even though the former uses the latter. Like the majority of nineteenth century natural scientists, Lodge thought that one should understand physical phenomena dynamically. Thus, he confided to Larmor in a letter of 9 September 1922:

> "Mathematical analysis rightly leaves the world behind and soars into a region of its own. A writer in The Literary Supplement to *The Times* has recently been saying that music does the same, and rather luminously remarks that this is the reason why one can have musical and mathematical prodigies in infants without the possibility of actual experience. Relativity confronts us with the attempt to link these mathematical soarings with physical fact. It is strange that physical phenomena allow themselves to be expressed in this way. But it cannot be the only way of expressing them. There must be a dynamical method too. But until we have evolved the dynamics of the ether we find that we can proceed by undynamical methods, just as [James] MacCullagh [(1809–1847), an Irish mathematician] found long ago."

In some sense, he saw himself and Larmor as missionaries in a strange land, the land of mathematical physics and mathematically expressed relativity, their obligation being to make physical sense of the theory: "The attempted marriage between genuine Physics and hyper-Geometry — he wrote also in the last mentioned letter — was sure to lead to confusions. They will doubtless straighten themselves out, largely perhaps by our aid. For I know no one else who can wield both weapons with equal ease."

---

[7] Here is what Buchwald (1981b: 374) said concerning Larmor's education at Cambridge: "In many respects Larmor was a typical, if uncommonly intelligent, product of the Cambridge-centred educational system in physics of the 1880s."

3.3. Debating about Relativity: Lodge, Jeans and Eddington

In spite of his mathematical lacunae and ignorance of the physical content of general relativity, Lodge did not mind entering in discussions with other people as to what the theory meant or what its consequences were. It is fortunate that he was that sort of man, for it helps us now to find out what other British scientists thought of relativity as early as 1917. James Jeans and Arthur Eddington were two of those scientists. Let us begin with Jeans.

Replying on 14 August 1917, to a letter of Lodge, Jeans stated:

"I do not regard relativity at all as a settled question but I feel it is much the most hopeful possibility. The only alternative seems to me to be the Fitzgerald contraction which in my opinion leads to endless difficulties. It seems almost to require all atomic motions to be governed by the old-fashioned electromagnetic equations. [...] indeed under them the atom would necessarily shrink and collapse, and they seem quite inadequate to account for line spectra etc. I feel reluctant to make such great sacrifices to keep light propagation through the old-fashioned ether, and then to accept a gravitation which is not propagated through the ether and a gravitational field which does not contract with motion through the ether.

I agree entirely about the complexity of the presentation of relativity.

Einstein is not a trained mathematician or I suspect he could have put the whole matter much better."

At this point, Jeans put forward a problem for considering gravitation along the lines of special relativity. Consider, he said, the equilibrium of the Sun moving with velocity v. Obviously, it is acted on by its own gravitation and by electromagnetic forces from its own electrons. If only the latter existed, then the Sun would contract into a $(1 - v^2/e^2)$ spheroid. "But — Jeans went on — if the gravitational spheroid is spherical, this prevents complete contraction. A compromise is effected between electromagnetic forces, striving for a Fitzgerald spheroid, and the gravitational forces striving for a strictly spherical shape. The result is a spheroid, less flat than the Fitzgerald one. From this difference

108 *Einstein's Relativity in Great Britain*

between contraction of the Sun and his measuring rod, an observer could determine it."

What Jeans seems to be saying is that if one has in mind Lorentz contraction plus classical (i.e., infinite speed of propagation) gravitation, then one should get a non-spherical geometry for the Sun, which would allow the determination of its motion through the ether. With Einstein's new theory (general relativity), he added, it was different:

> "Einstein suggests that a simpler scheme is got by adjusting the gravitational field also, so that everything contracts uniformly. In this case an observer can not measure v."

The difference between both approaches was that the former one "required two mechanisms of propagation, apparently unconnected; the latter [general relativity] virtually one."

It is not clear to what extent Jeans really understood then the meaning of Einstein's general theory; there are indications that he thought that general relativity was just special relativity plus gravitation, perhaps something like Poincaré's theory (Poincaré 1906). Indeed, in his August letter to Lodge, he pointed out that his analysis "does not agree in form with Einstein's presentation because of the different ways of measuring length." Apparently, Jeans was getting his information about general relativity from Willem de Sitter's articles (De Sitter 1916a, 1916b, 1917), although his opinion of them was not very good: "I fear — he confessed to Lodge — De Sitter has rather prejudged the reception of Einstein's theory by a too abstruse presentation."

Less than two years later, Jeans was still discussing relativity matters with Lodge, although on this occasion, the subject of the discussion was the special theory. According to Jeans (Jeans to Lodge, 21 March 1919):

> "There are two distinct questions:
> (A) The system of laws in which we believe
> (B) The mechanism which produces these laws.
> (A) is a question of observation which can be definitely settled in the laboratory; (B) is a question of speculation which probably cannot be finally settled at all. Indeed it seems to me that it is only a guess that

(A) may result from mechanism at all. We English being a practical race regard every theme as an engineering problem.

To my mind, if I see things rightly, relativity has only to do with (A). It is simply a guiding principle to suggest laws for trial, and so far these laws have always been confirmed by observation.

Question (B) is another matter. Relativity finds the laws, and we may then try to find a mechanism to account for them. My own feeling is that the ether is not likely to account for them and I doubt if any 'mechanism' in the engineering sense, can do so. But certainly I cannot prove this, and there can be as many opinions as men. I rather think Lorentz agrees with you and Einstein with me. I remember at Birmingham Lorentz wrote on the board (in the radiation discussion):

Matter-resonators-ether

and then said 'You see I have written *ether* so I believe in it still; those of you who do not may substitute the word 'vacuum': it will make no difference.' The last 5 words are of course the principle of relativity in a nutshell."

Jeans saw clearly Lodge's position, and, in the end, Lodge had to agree that relativity (special relativity, the theory he felt competent to judge) was a valuable help when trying to know how nature behaves; but what he could never accept is that his cherished ether was not, somehow, a more fundamental concept. However, he found some comfort in this regard because he thought relativity did not enter into conflict with ether. In this sense, he wrote to Edmund Whittaker (1 September 1922)[8]:

"I know that Relativity theory deals with Physics hyper-geometrically, and does not need ordinary mechanical or kinematical properties for their explication. It truly ignores the ether but by no means denies it. Nor should the possibility of arriving at results by other means prevent the attempt to arrive at them also in more dynamical fashion, with a better

---

[8] Lodge was prompted to make these remarks by the following comment of Whittaker (in a letter to Lodge, 30 August 1922): "I am not quite sure that Einstein would fully endorse what you say in the first line of page 8 [of Lodge's Silvanus Thomson Memorial Lecture]: in his Leyden lecture of 5 May 1920 he says 'The action of the theory of general relativity is a medium possessing no mechanical or kinematical properties.'"

110   *Einstein's Relativity in Great Britain*

understanding of what is really happening. Relativity, like the second law of Thermodynamics, and to some extent the conservation of energy, enables us to obtain results without clearly exhibiting the details of the process. There must be a dynamical theory, even if we are able to dodge it for a time; though it is true it does not follow that the dynamics of the ether is identical with the dynamics of ordinary matter. But the principle of least action, which is subterraneously the principle of Relativity, cannot be regarded as truly dynamical."

Although some specific problems were considered, Lodge's interchange with Jeans remained on an almost epistemological, rather general, level. It was different with the debate which Lodge carried on with Eddington in 1917.

Even though Lodge found general relativity too hard (i.e., too mathematical) to understand, he did not refrain from considering the sort of problems that Einstein's new theory had opened. He felt that his world, his classical and "ethereal" world, was in danger, and he did not hesitate to come to the fore in its defence. This he did in the pages of the *Philosophical Magazine* in 1917 and 1918, as well as corresponding with Arthur Eddington, who also used the pages of the same journal to answer Lodge's claims.

In his first paper, Lodge (1917a) analysed some astronomical consequences of what he called "the electrical theory of matter." As one would have expected of him, Lodge's starting points were expressions of the "electromagnetic worldview." Thus, he supposed the following: (1) the motion of matter through the ether has a definite meaning; (2) an extra inertia due to this motion is to be expected at high speeds, "in accordance with the Fitzgerald–Lorentz contraction"; and (3) this extra or high-speed inertia is not part of the mass, but is dependent on the ether and hence not subject to gravity. He thought that from these "reasonable hypotheses," all "quite independent of the theory of relativity," some astronomical consequences would follow, in particular, that "the outstanding discrepancy in the theory of the perihelion of Mercury would be accounted for by attributing a certain value to a component of the true solar motion through the aether in the direction of the planet's aphelion path." That is, he wanted to show that his 'electrical theory of matter' could also cope with astronomical problems, with the same success as general relativity. Thus, we read in his paper (Lodge 1917a):

"Professor Einstein's genius enabled him in 1915 to deduce astronomical and optical consequences (some not yet verified) from the principle of Relativity. [....] I wish to show that one of them [the anomalous motion of the perihelion of Mercury] at least can be deduced without reference to that principle."

Once the paper was published, Lodge sent it to Eddington, who thanked Sir Oliver (Eddington to Lodge, 2 August 1917):

"I was very glad to receive your Phil. Mag. paper and have been much interested in reading it.

The criticism that rises in my mind is that it seems unlikely that we can find any solar motion which will set Mercury right without upsetting something that is already accordant. Einstein's formula had the great advantage of having d where you have ed, consequently his correction to Venus & the Earth was negligible compared with observation, but I am afraid yours may not be."

Lodge's points aroused Eddington's interest quite rapidly, and four days later, he was writing to Lodge again on August 6: "After writing to you I worried out the matter to what seems a definite conclusion. I have written a paper for the Phil. Mag. and enclose it as I think you would like to see it." Indeed, Eddington asked Lodge, who was one of the editors of the journal, to communicate the paper, pointing out that he would "be extremely glad as it would remove any possibility of an appearance of hostility, which was, of course, far from my thoughts, for (although I am somewhat of an Einsteinian) I regarded the suggestion as particularly interesting."

In his paper, Eddington (1917a) began by pointing out that Lodge's explanation of the motion of the perihelion of Mercury was "comparatively simple, and on that account will be widely preferred to the recent theory of Einstein, which introduces very revolutionary conceptions, provided that it meets certain other astronomical requirements which seem necessary." To see if those "astronomical requirements" were really fulfilled, Eddington wanted to discover whether or not in removing the discordances for Mercury, Lodge's theory introduced discordances for Venus and the Earth, as he had suggested in his letter of August 2. "If the

112　*Einstein's Relativity in Great Britain*

explanation breaks down under this," he pointed out, "the discussion will make prominent a feature of the success of Einstein's theory which has perhaps not been sufficiently emphasized." Fortunately for general relativity, Eddington's results did not favour the electrical theory of matter, and the Plumian Professor of Astronomy at the University of Cambridge finally concluded politely that (Eddington 1917a: 167):

> "It is disappointing to find that this interesting suggestion, which gives a simple explanation of the most celebrated discordance of gravitational theory, is apparently unable to satisfy the most stringent test proposed."

However, the discussion did not end at this point. In the course of their correspondence, Lodge pointed out to Eddington that some basic equation the astronomer had been using required amendment when the mass was taken as variable. Eddington (1917b) took the criticism into consideration and looked into it, but he concluded in the October issue of *Philosophical Magazine* as follows: "So far as I can see, the conclusions of my previous paper are not materially modified by this more rigorous calculation."

Confronted with the technicalities displayed by Eddington, Lodge could but declare that, although he did not see how to disprove Eddington's conclusion, he felt "that the last word has not been said on the subject" (Lodge 1917b: 519). His own results for Mercury and Mars were too good "to be readily abandoned." He insisted on the hypothesis that the additional inertia due to motion is not part of the body's true mass, and so is not subject to gravity, offering a further argument (Lodge 1917b: 519):

> "In favour of the hypothesis of gravitation independence, I adduce the analogy admittedly not coercive-of a solid moving through a fluid. The apparent inertia of such a body is increased by an amount depending on the fluid displaced, but its floating or sinking properties remain unaffected: the extra inertia is not part of its mass, and is not subject to gravity. If the extra electrical inertia of moving matter is not part of the true mass, but represents only aetherial reactions, which is what I expect, then an astronomical perturbation is bound to be caused in rapidly moving planets; and whether this perturbation can be adjusted to agree with

observation, i.e., whether a solar drift can be chosen which shall give a result neither in excess for one planet nor in defect for another, becomes a matter for further detailed calculation."

On 8 December, Eddington received Lodge's new paper and wrote to him immediately:

"I was very pleased to receive the Phil. Mag. from you this morning, and to read your article, which gives a most excellent presentation of the position we have reached.

The point you now rise had not occurred to me before, but I feel no doubt it is quite true. If the extra-mass is subject to gravitation and the Newtonian law holds unmodified, the perturbations are just doubled. (At any rate the main perturbations-I have not examined those involving the eccentricity)."

However, no matter how "satisfactory" it might be "to have this point, on which there may be legitimate difference of opinion," Eddington thought that it should be put in the background because for him the real question was: "Does the strict Newtonian law hold good for systems moving through the aether?" And he added:

"In thinking over the matter in the last two months, I had felt that this discussion has produced valuable evidence (*evidence-not proof*) for the principle of relativity in its Minkowskian form, with or without the recent extensions of Einstein. We know that optical and electrical laws have entered into a strange conspiracy to prevent us determining our motion thróugh the aether, by methods which at first sight seemed almost certain of success. In generalising this, it has been assumed as an hypothesis (I believe without a vestige of proof) that gravitation conforms to the same principle. By working out your theory, we obtain the first definite indication that gravitation has actually joined the conspiracy. The observed variation of perihelia & eccentricities of Venus and the Earth are just what they would be if the sun were at rest in the aether; hence the Newtonian law has modified itself so as to conceal the effect of the sun's motion (which calculations lead us to expect to be easily observable), and we add one more to the long series of experiments giving null results."

Indeed, through his interchange with Lodge, Eddington's faith in relativity had been reinforced, while Lodge had been forced to retreat by lack of arguments.

3.4. Larmor and General Relativity

Joseph Larmor's first publication (Larmor 1919a) which is connected with general relativity appeared in 1919, that is, during "the year of the eclipse." Previously, he had told Lodge of his interest in, and difficulties with, Einstein's gravitational theory. Thus, on 10 March 1918, he wrote to his friend that having "been compelled by Indian young men and Newnham young women, the only audience, to lecture on relativity, of which they know as little as I do, I have hit upon the precise way in which both gravitational and kinetic mass must be altered on a planet or the Sun to secure it locally. If I could get settled I would contribute to your investigations-but this war!"[9] And a week later, he wrote again, now in a rather desperate mood (Larmor to Lodge, 18 March 1918):

"It is hopeless to *explain* their relativity. It does not mean that space and time are abolished; but that many kinds of space and time are undistinguishable because to any *set of equations* in one system there is a corresponding *set of equations* of like form in the other system — a covariant set, as the mathematicians say. You can't put that into words. There are a vast number of possible forms of relativity on this basis, short of David Hume's original form that we only know impressions on our organs of sense, and everything else is padding which can be filled in our pleasure. Einstein still keeps an external extension in

---

[9]The reference to "Indian young men and Newnham young women", not really a nice comment, should be understood as one of the consequences of the war, then in its last moments, which was responsible for a substantial reduction in the number of British males who followed courses at Cambridge University. Newnham was one of the two colleges for women which existed then in Cambridge (the other was Girton).

space & time as well as our sense organs; but in time he will doubtless eliminate them."

But in spite of such difficulties, he entered the "relativity arena." And his 1919 paper is, in one sense, an outstanding publication, as it has the merit of having introduced the first five-dimensional unified field theory in the history of the several efforts to find a unitary framework for the electromagnetic and gravitational interactions. The often mentioned five-dimensional unified theory developed by Kaluza (1921) was proposed two years after Larmor's, although it is true that it was more comprehensible than that of the Lucasian Professor of Cambridge.

Larmor's paper is, in several respects, rather chaotic, and the physical ideas behind it quite confuse, but it is nevertheless interesting, as it reveals some aspects of its author's relationship with Einstein's general relativity. However, initially, the paper was not related to this theory, but to the "electrodynamical scheme of relativity," or "electrodynamic relativity," that is, it was closer to special relativity. The idea was to use the symbolic calculus developed by the Irish Professor at the Department of Mathematics at Aberystwyth, William John Johnston (1858–1924), which, as Larmor (1919a: 336) points out, was a sort of reverse procedure of the one fol-lowed by Hermann Minkowski in his four-dimensional interpretation of special relativity:

> "The procedure of Minkowski seems to have been, having identified electrodynamic relativity with invariance of the system as regards posi-tion in the fourfold continuum, to group and identify the physical quanti-ties of the Maxwellian field as components of various 4-vectors and 6-vectors. [...] The procedure of the present calculus is the reverse. The system of invariants natural to a four-dimensional flat continuum are immediately manifest in Mr. Johnston's application of Clifford's calculus; and the totality of them are identified precisely with the vectors of the electrodynamic scheme of Maxwell, with which they are co-extensive."

Although Larmor was working essentially within the framework of electrodynamics, he was aware of the problem that this meant not having

116 *Einstein's Relativity in Great Britain*

included gravitation (Larmor 1919a: 342): "But, whatever be the critical obstacles, the problem of probable interaction between gravitation and electrodynamic fields, including rays of light, of course, remains urgent; and if such connection is actually detected by the various astronomical determinations now in progress, data such as hitherto have been entirely non-existent will have been supplied for an attack on this deep-seated question."

Larmor's paper had been received by the Royal Society on 28 August 1919,[10] but the previous comment was included in pages added on 20 October. In the meantime, the news of the results obtained by the British eclipse expedition began to arrive, and Larmor added a note to his statement: "This was written before it became known that the Greenwich and Cambridge astronomers, in their recent eclipse expeditions, had confirmed Einstein's prediction for the amount of the deflection of a ray of light by the influence of the Sun. It must be recognized that the theory has come to stay in some form or other. Its main implication, of instantaneous

---

[10] Shortly afterwards, on 12 September, during one of the sessions of the meeting of the British Association for the Advancement of Science held at Bournemouth (9–13 September), Johnston and Larmor had signed jointly a brief note on "The Limitations of Relativity." Among the points they stated there, figures the following, of an evident Newtonian tone (Gray 1920: 143–144): "If time were linked with space after the manner of the fourth dimension, relativity in electrodynamic fields would be secured as above, but the sources of the field could not be permanent particles or electrons. If physical science is to evolve on the basis of relations of permanent matter and its motions, time must be maintained distinct from space, and the effect of convection must continue to be thrown on to material observing system in the form of slight modification of its structure" (Johnston and Larmor 1920: 159). It is interesting to remember that during the presidential address of section A ("Mathematical and physical science") of that meeting, Andrew Gray, professor of Natural Philosophy at the University of Glasgow, voiced some misgivings toward relativity: "I have attacked Minkowski's paper more than once, but have felt repelled, not by the difficulties of his analysis, but by that of marshalling and keeping track of all his results. [...] Some relativists would abolish the ether,-I hope they will not be successful. I am convinced that the whole subject requires much more consideration from the physical point of view than it has yet received from relativists."

propagation of change in the constitution of space,[11] seems to be avoidable only on a psychological point of view which would assert that a portion of space is existent only while attention is concentrated on it. It can be managed, however, by including the varying space in a uniform space of higher dimensions, just as the deformation of a two-dimensional surface can be visualized as a whole in uniform space of three dimensions."

This "accommodation" — in a "uniform space of higher dimensions" — of the general relativity world favoured by the results of the eclipse expedition was performed by Larmor in a new addition to his paper (this one dated 20 November, that is, after the 6 November joint meeting of the Royal Astronomical Society and Royal Society). It was there that he developed the five-dimensional unified theory mentioned above. His purpose was no thing other than to maintain the symbolic calculus *à la* Johnston he had developed in the previous pages. That calculus was based on a four-dimensional flat continuum (a "homaloid" in Larmor's terminology); but now that Einstein had included the "phenomenon of gravitation" in that Minkowskian scheme "by altering slightly" the expression of the metric (of course, this is not a correct expression of the mathematical and physical structure of general relativity), Larmor (1919a: 353) thought that his previous ideas could still be maintained, and so, he pointed out that that "generalization can still be brought within the range of the Clifford geometry by introducing into the analysis a new dimension preferably of space." "Now," he continued (pp. 353–354), "any continuum of four dimensions, having a quadratic line-element, however complex, is expressible as a hypersurface in this homaloid continuum of five dimensions."

Like many others, in Britain and elsewhere, Larmor seems to have been stimulated by the general atmosphere aroused by the results of the eclipse expedition. He sent, for instance, a communication (Larmor 1919b) to one of the two main meetings organized in Great Britain soon after the one held on 6 November to discuss Einstein's 1915 theory: the session hosted by the Royal Astronomical Society on 12 December? As

---

[11] This is, of course, a blatant misunderstanding of general relativity, which reveals some of the limitations in Larmor's capacity to understand Einstein's gravity theory.

118 *Einstein's Relativity in Great Britain*

one could have expected, Larmor expressed again on that occasion his doubts concerning Einstein's theory, of which he said: "How far it is from being a determinate theory of the universe will appear to anyone who dips into the writings of its developers."

He might have been critical, but not indifferent. Thus, during the following decade, roughly until the late 1920s, Larmor's main fight in the relativity arena was to interpret Einstein's relativistic contributions "rightly," developing, if possible, a gravitational theory of his own, as in August, 12, 1922, when he wrote to Lodge from his summer house in Dhu Varenn, Portrush:

> "I thought I was going to write out my theory of Gravitation; but the sea air is too strong and sleepy for me. Moreover it is off colour. Gravitation is not an essential property of the Electron as I thought. In fact there is no answer to the fundamental question, why should there be Gravitation at all? Only if there is, considerations of isotropic form and symmetry restrict it to the Einstein field-form, if it is to fit in with the null effects of uniform translation (Michelson). Thus there is no explanation, but only a definite restriction of possibilities."

In September, he was still in a depressed mood, still trying to arrive at a satisfactory — satisfactory to him — interpretation of general relativity: "I have been worrying in the Einstein mazes. Why were they ever invented? So long as you stick to Algebra and don't try to interpret it you can sail away — goodness knows where. But to find out how it relates to external physical science is the rub". (Larmor to Lodge 1 September 1922).

It was a problem of two worlds, the new, relativistic one, and the old Newtonian and Maxwellian, being mixed up. Larmor tried to follow Einstein's general relativity arguments, but his world was a peculiar mixture of the Newtonian and Maxwellian worlds. And so he found problems in that strange land:

> "But in the fourfold the analyst hunts tensors and leaves the world behind. There you may interpret the world in the fourfold frame: you can also do it in the Newtonian frame-either frame is adequate for that

content, though they are not equally natural and coincident always. But when you mix the two, and use the developments of Newtonian dynamics in a fourfold pseudo-spatial frame, it is confusion of tongues.

The Maxwell stress is the mother of tensors. Its inner (very remarkable) signification is a most elusive business: I have thought I had it pinned down many a time for years past-and indeed have in a way. It is a case of one medium existing in another-matter in aether-and both in a frame of reference, and the interrelations take some disentangling. I fancy I see it all clear, but may not do so next week." (Larmor to Lodge 1 September 1922).

One of the problems Larmor tried hard to solve was the interaction between gravitation and light, a problem especially attractive to those who wanted to reproduce the deflection of light effect predicted by general relativity from a theory based on classical electrodynamics. He tried in 1920 (Larmor 1920) and tried again during 1923?[12] From his summer house, he wrote to Lodge on 23 August 1923:

"I have been here two months-trying to annex what of good there can be in Einstein. In his latest expositions there is one sensible thing latent, viz. Newton's absolute time, though he does not seem to know it.[13] Generally the process of 'saving one's face' seems to have begun.

---

[12] The following sentences from Larmor 1920: 333 contain some of the main ideas behind his efforts on this direction: "But we now pass from kinematic discussion of frames of reference to physical considerations. If we are to assert, in agreement with the doctrine of relativity plus Least Action, that inertia is a property of organised energy and proportional to it, therefore not solely of matter, and if we are to admit with Einstein, in the same and other connexions, that light is made up of small discrete bundles or quanta of energy, it would appear that each bundle is subjected to gravitation. Therefore if a bundle comes on from infinite distance [...] it will swing round the Sun in a concave hyperbolic orbit, and as the result, the direction of its motion will suffer deflection away from the Sun by half the amount that has been astronomically observed."

[13] Here, of course, was another of Larmor's problems with relativity. To Lodge, he wrote on 30 September 1925: "In the fourfold continuum, which is a natural geometric theme, there is no time nor motion nor direction; it is there pure length. Succession in time comes to the observer through vis ion by successively (directionless) rays."

120 *Einstein's Relativity in Great Britain*

I can include a propagated gravitation quite comfortably were it not for the now demonstrated interaction of light and gravitation, which is a fundamental experimental result."

In 11 October, he thought he had found the solution:

"I have I think fathomed the deflection of light, after many trials, (at any rate to my present satisfaction,) expressed as a result of bringing gravitation, taken as we find it but not explained, within the ambit of the electrodynamic relativity as confirmed by experiment.

The universe is full of absolute atoms, which on account of local relativity each tick out through their spectra the same absolute time to observers at rest relative to them. These local times are connected up physically into a universal time, for the absolute time a ray takes to pass, from a point P to another Q in the aether, is proportional to its length $P\,Q$, whether straight or curved. This universal absolute time T in the astronomical time postulated by Newton is no more unintelligible than absolute vibrating atoms identical everywhere in the universe.

But the aether-field is locally disturbed in constitution by adjacent large masses, as the deflection of light shows. If the disturbance is such that the equations of vibrations (and of electrodynamics generally) can be reduced back to the simple form by changing from astronomical time T to an auxiliary time-variable t, namely

$$dT = dt\,(1 - v/c^2)$$

depending on locality, the difference being sensible only near large areas, the deflection of light will be as observed, and all the demands of local-relativity as observed will be met, provided this transformation is accompanied by one representing slight radial warping of the space of influence near each mass.

But this — a fixed ether except as locally affected by masses moving in it, and resulting universal atomic time —, is of course anathema to the relativist philosophers. The universe can however curl up into a ball if they so wish it."

One paper, entitled "On the nature and amount of the gravitational deflection of light" (Larmor 1923a), was the result of those efforts. However, he came to realize that its content was unsatisfactory:

Larmor, Lodge and General Relativity 121

"I am worried over adapting a scheme of gravitation to the experimental deviations of rays and spectral lines (if they are final) now for two years without success. But what I am absolutely certain about, however I turn it over, is that Einstein's theory gives only half results. (As in Phil. Mag. Dec. 1923)" (Larmor to Lodge, 30 September 1925).

However, he kept studying Einstein's works, perhaps with more interest and attention than most British scientists at the time. By the end of 1924, it appears he had restricted his main differences with Einstein to points of interpretation. He did not accept that it was a "final theory" (whatever this may mean), a satisfactorily physical explanation of nature, but he began to appreciate some of its values as well as Einstein's powers:

"I have been reading Einstein several times over-and begin to fathom his mentality. It is most picturesque and acute: he gets to where he wants without great regard for coherence. [...] He never gives any references (except to his patron Lorentz,) or mentions the people who must be in his mind in his "reflections".
I agree (with Lorentz) that he is a very powerful phenomenon: his "reflections" on quanta of light etc. are much in the same boat-not a logical system at all, but what he adopts as the best choice between alternatives all imperfect; and the choice once made, his business is to bolster it up the most plausible way he can. Is it the oriental imaginative mind of [*sic*; at?] play? Ask any Anglo-Indian?"[14] (Larmor to Lodge, 9 September 1924)

Even so, he could not avoid entering into complex and obscure disquisitions:

"The spectrum test is *really* a test of absolute intervals of time! As such I accept its results. The frame of reference is *not* of no consequence. Every system has its absolute frame; namely the frame convected with the system as a whole. Thus each atom has its own frame, each solar

---

[14] Perhaps, we should consider the last sentences as an indication of anti-Semitism in Larmor.

122   *Einstein's Relativity in Great Britain*

system has its own frame; intervals of time in these frames are absolute, because the atoms are such. The function of relativity is to piece them together into a universal frame. If it is to be a space-like extension, this strictly cannot be done. We must therefore discuss *all* cosmic history as one unit, as pieced out by memory, which is astrophysics. The defect of the gravitation scheme is that time in it is entirely illusory; it is not possible to divide the cosmic history into slices so that the following ones are determined by the prior ones. There could be no causation, *no evolution* of the cosmos, if gravitation were 10^4 times as intense as it is, etc. Yet there it is, in possession.

The fundamental fact is that you exist and that you are intellectually identical with me; and I to verify that. Intellectual atomic theory.

You see I am bitten by the metaphysic mosquito."

3.5. Larmor, General Relativity and the Least Action Principle

In Chapter 1, I mentioned Larmor's attachment to the Least Action Principle, now I come back to this point.

According to Eddington (1942–1944: 204), Larmor became convinced of the truth of general relativity around 1924; at that time, he said to him: "I have been reading continental writers on relativity, and I find it is all least action. I begin to see now." However, his letters to Lodge do not give quite the same impression. Not that he did not emphasize the role of his beloved Least Action Principle. On the contrary, he thought that there laid the real truth, but that general relativity did not incorporate it in a correct manner. Thus, he told Lodge as late as 30 September, 1930: "Eddington is facetious about the Universe being governed by least action. But the Einstein geodetic orbit is only a case of Least Action *wrongly applied,* as I feel certain. From Vienna they admit to me that there is something wrong." And he added: "The view seems to be that the Mach (sensational) philosophy which fascinates Einstein is so inevitable that innate contradictions are merely provisional imperfections."

These sentences seem to indicate that Larmor had some philosophical interests and knowledge of Ernst Mach's works (certainly of the influence that Mach's ideas exerted on Einstein during part of his career), and perhaps also of the Vienna Circle contributions, although one of the expressions he used, "they admit to me that there is something wrong," makes this last possibility rather improbable, in as much as most of the members of the Vienna Circle (in particular, Moritz Schlick and Rudolf Carnap) were sympathetic to Einstein's approaches and theories. Further evidence would be needed, however, to understand the real meaning of those sentences of Larmor as well as his philosophical interests, if any.

What is clear is that he tried to use Least Action as a tool, or help, to undermine general relativity, as well as to develop his own "physical relativity." As when, late in 1922 and early in 1923, he found attractive the criticisms of general relativity made by the French physicist Jean Le Roux (1922a, 1922b, 1922c), which led him to think "that the Einstein *philosophy* is near its end" (Larmor to Lodge 12 December 1922).[15] A month later, 10 January 1923, he wrote to Lodge in this same sense:

> "I have finished my proof that though the Le Roux paradox damns all Einsteinism beyond repair, it leaves my Action in a medium alone. You must be educated into Action. See my edition of Maxwell's 'Matter and Motion' [1920]."

Le Roux's criticisms of Einstein's general relativity were as radical as obscure, offspring of a mind that mingled together different worlds and concepts — those of Newton's mechanics, Euclidean and Riemann's geometries, and Einstein's gravitational theory — and that could not understand the essence of the approximation procedures which connect in general relativity those worlds and concepts (not to mention wrong assumptions). The sort of problems which were also responsible

Specifically, what Le Roux argued against general relativity was the following:

(1) To arrive at the notion of curvature of surfaces, it is necessary to consider simultaneously two quadratic forms, one of them having a fixed

---

[15] Lodge was also attracted by the Frenchman's work. In a letter he wrote to Larmor on 13 January 1923, he mentioned that he had already read "the first of Le Roux's papers in the C[omptes] R[endus]" and that he was "much tickled with it."

physical meaning. As in the general theory of relativity, Le Roux went on, one considers only one form, the invariants derived from it cannot be associated to properties of space; thus, the question of whether the space(time) is finite or infinite cannot be answered (Le Roux 1922a).

(2) The success of general relativity in explaining effects like the perihelial shift of Mercury is only apparent, as Einstein's theory does not take into account the perturbations due to mutual actions between the planets of the solar system; and if one suppresses such perturbations, "the concordance disappears," leaving an unexplained shift of 531" (Le Roux 1922b).

(3) Newton's mechanics is not an approximation of Einstein's theory because both "are based on completely different principles" (Le Roux 1922c).

The fact that a physicist like Larmor was attracted almost immediately by such ideas (Le Roux's papers were communicated to the Académie des Sciences in November and December 1922), due to a little-known French scientist, shows, perhaps better than in any other way, how prone he was to react in favour of arguments against relativity, or, what is the same, how much he disliked general relativity.

Coming back to the Least Action principle, we have seen that Lodge was not as impressed with the principle as was his friend, as can be seen from a letter he sent Larmor on 9 January 1923:

"The present fashionable mode of writing down equations without specifically referring to an ether, or to any form of mechanism, leaves everything in the air, and prevents the formation of clear conceptions, and evades everything of the nature of explanation or real understanding.

I feel something of the same sort in connexion with your Action theories.

They are no doubt powerful and necessary; but none of these main force methods-like the Second Law of Thermodynamics, for instance, or even the Conservation of Energy-can be the last word. They enable one to arrive at a result, but they don't explain it. And until something like a real explanation is given, the ether will not come into its own."

Finally, Eddington's assertion is misleading in another sense. When he put on Larmor's lips the words: "I have been reading the continental writers and I find it is all least action," one might think that the Lucasian Professor had found Least Action in the works of those continental writers. Read, however, what Larmor wrote to Lodge on 1 September 1922:

"None of our foreign savants nowadays know what Hamilton's principle of Varying Action means. They talk glibly of the Hamiltonian principle as if it were a thing to play with at random, instead of the precise kernel of all dynamics. If only they would read their own Helmholtz if they can't read English. I suppose Jacobi had something to do with perverting it off the physical track into general algebraic analysis."

What is certainly true of what Eddington pointed out in his obituary of Larmor is that as I have remarked above, Larmor "wavered very much over Einstein's theory of gravitation" (Eddington 1942–1944: 205).

"This is one of the reasons why his papers dealing with — or related to — general relativity are so difficult to understand; one never knows what he was aiming al. In the end, he came to value so little his "physical relativity" papers that he decided not to include them in the edition he prepared of his *Mathematical and Physical Papers?*"[16]

Explaining this omission in the preface he added to volume II, Larmor (1929b: vii) referred to general relativity in the following terms:

"In this auxiliary cosmos space and time and motion do not occur; yet it can be of great value in a geographical sense, after the manner of a spherical map of the Earth, for unravelling the intricacies of relations connecting regions in our actual world, which astronomers are permitted to formulate only in terms of the complications of delay in the messages of the informing rays of light. The parallelism of relations between the world of actual perceptions and memory and this particular

---

[16] Apart from those already mentioned in the preceding pages, these papers are Larmor (1921, 1923b, and 1927).

conglomerated fourfold assembly of permanent ray-models, so to say, extends over a prominent yet necessarily limited range whose boundaries have hardly yet been very closely examined. There may be, however, people who aim at transferring their whole life into this new cosmos. The papers now omitted are concerned largely with various general aspects of this correlation: naturally much of their contents has now become transcended, or modified into improved presentations."

To which "improved presentations" Larmor was referring in the last quotation is not really worth asking; what is clear is that when he wrote these sentences, relativity had, if not convinced, at least begun to win him over. Were we to think in terms of Lakatos's scientific research programs (Lakatos 1970), we would conclude that the Newtonian–Maxwellian research program that Larmor (and Lodge) tried to develop had finally arrived at a clearly degenerative stage, the progressiveness lying on the side of the Einstenian research program. Ever since he considered that the special and general theories of relativity had won enough attention and consideration from the scientific community, Larmor had tried to criticise their most novel concepts simply because it was not possible to understand, or reduce, them to classical, i.e., Newtonian and Maxwellian, terms ("The main difficulty about the acceptance of the relativity theory — Max Planck once wrote — was not merely a question of its objective merits but rather the question of how far it would upset the Newtonian structure of theoretical dynamics" [Planck (1930); English translation: Planck (1932: 44)].) At the same time, Larmor, or, again, Lodge, had kept trying to accommodate results emanating from the relativistic theories to the framework of classical Newtonian and Maxwellian physics. He, they, the "old guard', were always one step behind the facts and developments; their efforts were not to uncover a new phenomenal or conceptual world, but to accommodate the world that relativity had discovered — or was still discovering — in the physics they loved, the physics of Newton and Maxwell. Not only were they unable to accomplish such a program; while pursing it, they were constantly losing the simplicity and coherence that their beloved physics had originally had. No doubt, finally, Larmor understood this, and so abandoned his efforts. His decision to exclude his "physical relativity" articles from his published *Papers* constitutes clear evidence in this sense.

Einstein with Robert Haldane, during the visit of the physicist to England. Photograph by Walter Benington, published in *The Sphere* on 18 June 1921.

Chapter 4

The Early Reception of Relativity Among British Philosophers, or the Ductility of Philosophy

"I've become very good friends with the Ehrenfest's children and play with them very much. I also have to study Cassirer's manuscript, which is less amusing. These philosophers are peculiar birds."

Albert Einstein (1920).[1]

"The theory of relativity is primarily a physical theory. But anyone who wishes on that account (as has sometimes been done) to deny the philosophical character and scope of the theory, has failed to realize that the physical and philosophical viewpoints can by no means always be rigorously separated from each other; that each, on the contrary, passes into the other as soon as they address themselves to dealing with the highest and most general concepts of physics."

Moritz Schlick (1922).[2]

---

[1] Albert Einstein to Elsa Einstein, 19 May 1920, reproduced in CPAE (2006: 264–265). For the English translations, I have used the versions included in the English edition of this volume: Henstchel, transl. (2006: 164).

[2] English translation in Schlick (1979c: 343).

130 *Einstein's Relativity in Great Britain*

4.1. Introduction: Relativity and Philosophy, its German Roots

Relativity, both the special and the general theories, has maintained an intimate relationship with philosophy, thousand of pages have been written dealing with its philosophical implications. More problematic is both the geographical distribution (i.e., the nationalities) and the ideas of those who wrote those pages.[3] Were they mainly German-speaking individuals, people like Schlick, Carnap, Reichenbach, Cassirer or Popper? Were they scientists turned philosophers (briefly — like, for example, Poincaré, Mach, Eddington or Bridgmann — or more permanently), or philosophers who recognized the fecundity and relevance to their discipline of Einstein's contributions (as was in the cases of Bergson and Russell)?

I will not attempt to answer in this chapter these questions in all their generality, but only to provide some details which can help in producing a broader panorama. Thus, I will concentrate on the early reception of Einstein's relativity among British philosophers. My study suggests that it is not true what wrote John Passmore (1966: 334) in his splendid, although partial ("it is written," confess its author in the preface, "from an English point of view"), book *100 Years of Philosophy*[4]:

> "Professional philosophers, as distinct from philosophical journalists, have been singularly little affected by the revolution in physics. They have been inclined to suspect that, like a great many other revolutions, the revolution in physics raised no new philosophical problems and settled no old ones, for all the dust and fury. As well, it must be confessed, professional philosophers have been intimidated by the mathematics into which philosophical physicists so gratefully sink at crucial points of their reasoning; nor has the philosophical crudity of what they could

---

[3]Contributions to these questions are Ryckman (2005), Hentschel (1990), for the Italian case Maiocchi (1985) as well as Reeves (1987: 206–208), for the Soviet reaction Graham (1972) and Vucinich (2001); though not dedicated exclusively to the philosophers' reaction, some interesting details are contained for the French case in Biezunski (1981) and Paty (1987), as well as in Biezunski, ed. (1989), where the correspondence that Einstein maintained with the philosopher Émile Meyerson and the scientist interested in philosophy André Metz is reproduced.

[4]Valuable information concerning many of the British philosophers considered in this chapter appears in Metz (1938). See also, Krauss and Laino, eds. (2023).

The Early Reception of Relativity 131

understand led philosophers to expect any considerable degree of illumination from what passes their comprehension."

It is true, however, that immediately after having said so, Passmore debilitates his previous affirmation by admitting that: "There are, of course, exceptions. That very remarkable philosopher-statesman, R.B. Haldane, in his widely read *The Reign of Relativity* (1921), attempted to incorporate Einstein's theory within the Hegelianism to which he so faithfully adhered; Alexander welcomed what he took to be a partial confirmation of his theory of space-time; Russell wrote popular expositions of the new physics and shows traces of its influence; and a number of Cambridge philosophers, such a C.D. Broad coped manfully with the attempt to make philosophical sense out of contemporary developments in physics." But I will say more later on about Robert Haldane.

How many and how rapidly or slowly philosophers reacted as regards relativity is important, but there are other questions that deserved to be considered. One of them is the nature of their reaction; were they favorable or not to the new physical theory? Again, of course, the answer to such question might perhaps be different in different countries or philosophical communities. Thus, it seems that the German case was rather specific, as a glimpse of Einstein's already published correspondence suggests. Let us see a few examples in this sense, before going to the British case.

The appearance of philosophers as Einstein's correspondents began especially in 1919, the year of the British eclipse expedition that made him a world celebrity. Einstein's already available correspondence is a good proof of this, although in the German case we have ample evidence of interests by philosophers before the announcement (November) of the results of the expedition. So, on 3 May 1919, Einstein wrote to the philosopher Hans Vaihinger, a defender of philosophical relativism[5]:

"I do see that Study does not do you justice. I gave you the book only because it is written with such wit and is amusing, not because I wanted to plead for that view. I find that his 'realism' is philosophically quite

---

[5] CPAE (2004: 51), Hentschel, transl. (2004: 29–30). Vaihinger (1852–1933) was emeritus Professor of Philosophy at the University of Halle and founding editor of the journal *Kant-Studien*. He was the author of a book, *Die Philosophie des Als Ob* (*The Philosophy of As If*), that Moritz Schlick quoted favorably often in his *Allgemeine Erkenntnislehre*.

132   *Einstein's Relativity in Great Britain*

foggy, only substantiated in the little book so that the reader gets a Molière-like slap. I find the book *Allgemeine Erkenntnislehre* [*General Theory of Knowledge*] by M. Schlick, which takes a somewhat similar point of view, much more penetrating...

The assumption that I held out the prospect of a paper for the *Kant-Studien* is based on error. I am too little versed in philosophy to take an active part in it myself; if I can be passively receptive to the work of the man in this field, I am content enough. I just promised to pass on information, verbally as well as in writing, about matters regarding my specialty in particular that are of interest to philosophy. This is the only way in which I can perhaps be of service to philosophy. Cobbler, stick to your last!"

The mentioned Moritz Schlick (1882–1936) was one of the earliest and more active "missionaries" of Einstein's relativity in the philosophical world.[6] A student of Max Planck, under whom he got his Ph.D. in physics in 1904, with a thesis on the reflection of light in inhomogeneous media, Schlick turned afterwards to philosophy and was soon attracted by the many wonders of relativity, as can be seen, for example, in his 1915 article on "Die philosophische Bedeutung des Relativitätsprinzip" ("The philosophical significance of the principle of relativity;" Schlick 1915), or in his rather general exposition of Einstein's relativity theories, the special and the general, that appeared in 1917, in two parts, in the scientific weekly journal *Die Naturwissenschaften*, as well as, expanded, in book format (Schlick 1917a, b). But they were not these works the ones that made him famous in the philosophical world (not his previous book, *Lebensweisheit: Versuch einer Glückseligkeitslehre*; 1908), but his *Allgemeine Erkenntnislehre* (*General Theory of Knowledge*), whose first edition appeared in 1918, with a second and considerably revised edition in 1925. Of this book, Alfred Ayer (1987: 122–123) wrote: "In his *General Theory of Knowledge* [Schlick] insisted that every scientific statement or

---

[6]According to Herbert Feigl (1937/38, 1979: xix), Schlick's "hindsight into scientific questions was quite specially revealed when, as early as 1915 (long before general attention was directed to Einstein), he became possibly the first philosopher to publish a high-grade epistemological analysis of the theory of relativity." The work in question was Schlick (1915).

theory must be capable of being verified, in the sense that it had to have consequences which were capable of corresponding to observable facts, but the observable facts could have physical objects for their constituents. He agreed with Mach in rejecting psycho-physical dualism, arguing that to speak in mental or in physical terms was just to adopt one or other way of describing the same phenomena, but he tended to treat the phenomena as physical, in some degree anticipating the current fashion of identifying mental occurrences with processes in the central nervous system. This was another view which he was later to revise in favour of neutral monism. Perhaps the most remarkable feature of Schlick's book was that he anticipated Wittgenstein, with whom he had not yet made contact, in rejecting Kant's view that there could be such things as synthetic *a priori* truths and holding that all true *a priori* propositions, such as those of logic and pure mathematics, were analytic, or, in other words, tautological."

Indeed, *Allgemeine Erkenntnislehre* was a general philosophical treatise, not concerned specifically with Einstein's theories of relativity, though one question related to it — the correctness or incorrectness of Kant's ideas concerning Euclidean space as an *a priori* concept — was of special interest to Schlick (1972: 354):

"Euclidean geometry has served as the geometry of everyday life, and until a short time ago it seemed to provide the proper foundation for all the purposes of natural science. The new physics, however, in one of its boldest and most beautiful moves, has concluded from the Einsteinian Theory of Gravitation that we cannot make do with the Euclidean metrical determination if we wish to describe nature with the greatest accuracy and by means of the simplest laws. According to this theory, a different geometry must be used at each place in the world, a geometry that depends on the physical state (the gravitational potential) at that place. On the basis of Einstein's latest work it is likely that world space as a whole can best be viewed as endowed with approximately 'spherical' properties (thus as finite, although of course also unbounded).

It cannot be emphasized too much that we are not *compelled* to conceive of space in accordance with a theory of this kind. No experience can prevent us from retaining Euclidean geometry if we insist on doing so. But then we do not obtain the simplest formulations of the laws of nature, and the system of physics as such becomes less satisfactory.

134   *Einstein's Relativity in Great Britain*

Nevertheless, anyone who has been preoccupied with Einstein's theory and has come to know its inclusiveness, which simplifies the entire world picture so magnificently, will no doubt that the monopoly of Euclidean geometry in physics is at an end. The physical description of nature is not tied to any particular geometry and no intuition dictates that we must base such a description on the Euclidean axiom system as the only correct one, nor, of course, on any of the non-Euclidean systems either. We select — in the beginning instinctively, in more recent times deliberately — those axioms that lead to the simplest physical laws. In principle, however, we could have chosen other axioms if we were willing to pay the price of more complicated formulations of the laws of nature. Thus fundamentally the choice of axioms is left to our discretion."

Einstein was particularly attracted by Schlick's ideas. Thus, on 19 April 1920, he wrote to him[7]: "Your epistemology has made many friends. Even Cassirer had some works of acknowledgment for it; only the droning pastoral Liebert not. Young Reichenbach has written a very interesting paper about Kant & general relativity, in which he also gives your comparison with a calculating machine."

We find in this excerpt the names of two German-speaking philosophers who wrote extensively about relativity: Ernst Cassirer (1874–1945) and Hans Reichenbach (1891–1953). In due time, by the way, both left Germany and made the United States their country (both as professors of philosophy: Cassirer at Yale University since 1932, and Reichenbach at the University of California, Los Angeles, since 1938).[8]

Cassirer, who had grown up philosophically as a member of the neo-Kantian school of Marburg, was one of the philosophers who recognized that they had to revise their philosophical views so as to see whether they were consistent or not with Einstein's relativity theories. In his case, he was particularly interested in finding out whether the philosophical world-view that he had presented in his *Substanzbegriff und Funktionsbegriff* (*Substance and Function*; 1910), which was dominated by the Newtonian conceptions of space and time, was consistent with new relativity world.

---

[7] CPAE (2004: 510), Hentschel, transl. (2004: 317).

[8] Interesting comments about the relationship between Einstein, Cassirer, Reichenbach and also Schlick are included in Howard (1992).

Thus, after producing a manuscript about it and having sent it to Einstein, he wrote him on 10 May 1920, thanking for his help[9]: "Please accept my cordial thanks for your kind willingness to glance briefly through my manuscript now. [...] As far as the content of my text is concerned, it evidently does not propose to list all philosophical problems contained in the theory of relativity, let alone to solve them. I just wanted to try to stimulate general philosophical discussion and to open the flow of arguments and, if possible, to define a specific methodological direction. Above all, I would wish, as it were, to confront physicists and philosophers with the problems of relativity theory and bring about agreement between them."[10]

The manuscript in question was published next year in 1921, as a book entitled *Zur Einstein'schen Relativitätastheorie (Einstein's Theory*

---

[9] CPAE. (2006: 255), Hentschel, transl. (2006: 158).

[10] Although somewhat besides the purposes of the present book, it might be interesting to mention the comments that Einstein made (5 June 1920) concerning Cassirer's manuscript: "Highly esteemed Colleague, I studied your treatise thoroughly and with very much interest and admired, above all, how securely you master the essence of relativity theory. I made brief comments in the margin where I was not completely in agreement. e.g., I could not accept your opinion about the Kant–Newton relationship with the reference to space and time. Newton's theory requires an absolute (objective) space in order to be able to attribute real meaning to acceleration, which Kant does not seem to have recognized. I can understand your idealistic way of thinking about space and time and also believe that one can thereby arrive at a consistent point of view. Not being a philosopher, the philosophical antitheses seem to me more conflicts of emphasis than fundamental contradictions. What Mach calls *connections* [*Verknüpfung*] are for you the ideal names that make experience possible in the first place. You, however, emphasize this aspect of knowledge, whereas Mach wants to have it appear as insignificant as possible. I acknowledge that one must approach experiences with some sort of conceptual tool in order for science to be possible; but I do not think that our choice of these tools is constrained *by virtue of the nature of our intellect*. Systems of concepts seem empty to me, if the way in which they are to be related to experience is not laid down. This seems to me highly essential, even though we often find advantage in theoretically isolating purely conceptual relations, in order to have the *logically* secured interdependencies come more cleanly to the fore. With the interpretation of *ds* as a result of measurement that can be obtained in a very specific way by means of measuring rods and clocks, the theory of relativity stands and falls as a *physical* theory. I think that your treatise is very well suited to clarify philosophers' ideas and knowledge about the physical problems of relativity." CPAE (2006: 293–294), Hentschel, transl. (2006: 182).

136 *Einstein's Relativity in Great Britain*

of Relativity). "The following essay — Cassirer (1953: 349) wrote in the Preface — does not claim to give a complete account of the philosophical problems raised by the theory of relativity. I am aware that the new problems presented to the general criticism of knowledge by this theory can only be solved by the gradual work of both physicists and philosophers; here I am merely concerned with beginning this work, with stimulating discussion, and, where possible, guiding it into definite methodic paths, in contrast to the uncertainty of judgment which still reigns."

I will not intent analyzing Cassirer's ideas as expressed in this work. Let me, however, mention that he was not so much impressed by the new scientific theory as to assume now that the aims of physics and philosophy were the same. In this sense, it is worth quoting the final paragraphs of his book, which follow a few comments about the different meaning space and, especially, time have in disciplines such as history, painting or music (Cassirer 1953: 456):

"What space and time truly *are* in the philosophical sense would be determined if we succeeded in surveying completely this wealth of nuances of intellectual meaning and in assuring ourselves of the underlying formal law under which they stand and which they obey. The theory of relativity cannot claim to bring this philosophical problem to its solution; for, by its development and scientific tendency from the beginning, it is limited to a definite particular motive of the concepts of space and time. As a physical theory it merely develops the meaning that space and time possess in the system of our empirical and physical measurements. In this sense, final judgment on it belongs exclusively to physics. In the course of its history, physics will have to decide whether the world-picture of the theory of relativity is securely founded theoretically and whether it finds complete experimental verification. Its decision on this, epistemology cannot anticipate; but even now it can thankfully receive the new impetus which this theory has given the general doctrine of the principles of physics."

As to Reichenbach, he became interested in the theory of relativity before the news of the eclipse expedition made Einstein and his theories world famous. "Due to my work in the army radio troops [signal corps]," he wrote in an autobiographical sketch (Reichenbach 1978: 2), "I became involved with radio technology and during the last year of the war

The Early Reception of Relativity 137

[World War I], after I was transferred from active duty because of a severe illness I had contracted at the Russian front, I began to work as an engineer for a Berlin firm specializing in radio technology (from 1917 until 1920). During this period, and in my capacity as physicist, I directed the loud-speaker laboratory of this firm. I also got married. Soon thereafter, my father died and for the time being I could not give up my engineering position because I had to earn a salary in order to provide for my wife and myself. Nevertheless, in my spare time, I studied the theory of relativity; I attended Einstein's lectures at the University of Berlin; at that time, his audience was very small because Einstein's name had not yet become known to a wider public. The theory of relativity impressed me immensely and led me into a conflict with Kant's philosophy. Einstein's critique of the space-time-problem made me realize that Kant's *a priori* concept was indeed untenable. I recorded the result of this profound inner change in a small book entitled *Relativitätstheorie und Erkenntnis Apriori* [1920]."[11]

---

[11] Another future great central European philosopher who was deeply influenced by what he learnt during his youth of Einstein's relativity was Karl Popper. As he wrote in his autobiography (Popper 1974: 28–29): "Looking back at that year [1919] I am amazed that so much can happen to one's intellectual development in so short a spell. For at the same time I learned about Einstein; and this became a dominant influence on my thinking — in the long run perhaps the most influence of all. In May 1919, Einstein eclipse predictions were successfully tested by two British expeditions. With these tests a new theory of gravitation and a new cosmology suddenly appeared, not just as a mere possibility, but as a real improvement on Newton — a better approximation to the truth. [...] I was fortunate in being introduced to these ideas by a brilliant young student of mathematics, Max Elstein, a friend who died in 1922 at the age of twenty-one. [...] He drew my attention to the fact that Einstein himself regarded it as one of the main arguments in favour of his theory that it yielded Newton's theory as a very good approximation; also, that Einstein, though convinced that his theory was a better approximation than Newton's, regarded his own theory merely as a step towards a still more general theory. [...] No doubt Einstein had all this, and specially his own theory, in mind when he wrote in another context: 'There could be no fairer destiny for any physical theory than that it should point the way to a more comprehensive theory, in which it lives as a limiting case.' But what impressed me most was Einstein's own clear statement that he would regard his theory as untenable if it should fail in certain tests... This, I felt, was the true scientific attitude. It was utterly different from the dogmatic attitude which claimed to find 'verifications' for its favourite theories. Thus I arrived, by the end of 1919, at the conclusion that the scientific attitude was the critical attitude, which did not look for verifications but for crucial tests; tests which could *refute* the theory tested, though they could never establish it."

138 *Einstein's Relativity in Great Britain*

As we shall see later on, most British philosophers tried to argue that their philosophical views agreed with Einstein's relativity theories. The previous quotation suggests that it was not so in Reichenbach's case, perhaps because his training was different: he had studied first engineering at the Technische Hochschule in Stuttgart from 1910 to 1911, and realizing that his interests were predominantly theoretical he switched to mathematics, physics and philosophies at the universities in Berlin, Munich and Göttingen, having as teachers scientists as Planck, Sommerfeld, Hilbert and Born, and the philosophers Ernst von Aster and Ernst Cassirer.[12] In other words: when he became acquainted with relativity he was still philosophically ductile.

It gives an idea of Reichenbach's intentions in what he wrote to Einstein (15 June 1920), when asking permission to dedicate him his book *Relativitätstheorie und Erkenntnis Apriori* (*The Theory of Relativity and A Priori Knowledge*)[13]: "You know that with this work my intention was to frame the philosophical consequences of your theory and to expose what great discoveries your physical theory have brought to epistemology... I know very well that very few among tenured philosophers have the faintest idea that your theory is a philosophical feat and that your physical conceptions contain more philosophy than all the multi-volume works by the epigones of the great Kant. Do, therefore, please allow me to express these thanks to you with this attempt to free the profound insights of Kantian philosophy from its contemporary trappings and to combine it with your discoveries within a single system." To this letter, Einstein replied (30 June 1920)[14]: "The value of the th. of rel. for philosophy seems to me to be that it exposed the dubiousness of certain concepts that even in philosophy were recognized as small change. Concepts are simply empty when they stop being firmly linked to experience."

Indeed, Reichenbach was a philosopher who Einstein respected all his live. He was, for instance, one of the contributors to the famous volume of The Library of Living Philosophers directed by Paul A. Schilpp, *Albert*

---

[12]For the differences in philosophical outlooks between Cassirer and Reichenbach, see Rotenstreich (1982: 187–189).

[13]CPAE (2006: 313–314), Hentschel, transl. (2006: 195).

[14]CPAE. (2006: 323), Hentschel, transl. (2006: 201).

Einstein: Philosopher-Scientist. In his article there, Reichenbach (1949: 290) resumed splendidly the philosophical implications of relativity:

"it would be another mistake to believe that Einstein's theory is not a philosophical theory. This discovery of a physicist has radical consequences for the theory of knowledge. It compels us to revise certain traditional conceptions that have played an important part in the history of philosophy, and it offers solutions for certain questions which are as old as the history of philosophy and which could not be answered earlier. Plato's attempt to solve the problem of geometry by a theory of ideas, Kant's attempt to account for the nature of space and time by a '*reine Anschauung*' and by a transcendental philosophy, these represent answers to the very questions to which Einstein's theory has given a different answer at a later time. If Plato's and Kant's doctrines are philosophical theories, then Einstein's theory of relativity is a philosophical and not a merely physical matter. And the questions referred to are not of a secondary nature but of primary import for philosophy; that much is evident from the central position they occupy in the systems of Plato and Kant. These systems are untenable if Einstein's answer is put in the place of the answers given to the same questions by their authors; their foundations are shaken when space and time are not the revelations of an insight into a world of ideas, or of a vision grown from pure reason, which a philosophical apriorism claimed to have established. The analysis of knowledge has always been the basic issue of philosophy; and if knowledge in so fundamental domain as that of space and time is subject to revision, the implications of such criticism will involve the whole of philosophy."

It is interesting to emphasize Carnap's phrase, "it would be another mistake to believe that Einstein's theory is not a philosophical theory," as the question of whether relativity should be considered a physical or a philosophical theory was a recurrent theme in the philosophical debates.

Finally, I will mention Rudolf Carnap's doctoral dissertation, another example of philosophical early reaction to Einstein's space-time views (Carnap 1921). Here it is how Carnap (1963: 11–12) referred to this work of him: "In my doctoral dissertation, *Der Raum*, I tried to show that the contradictory theories concerning the nature of space, maintained by mathematicians, philosophers, and physicists, were due to the fact that

140 *Einstein's Relativity in Great Britain*

these writers talked about entirely different subjects while using the same term 'space.' I distinguished three meanings of this term, namely formal space, intuitive space, and physical space. Formal space is an abstract system constructed in mathematics, and more precisely in the logic of relations; therefore our knowledge of formal space is of a logical nature. Knowledge of intuitive space I regarded at that time, under the influence of Kant and the neo-Kantians, especially Natorp and Cassirer, as based on 'pure intuition' and independent of contingent experience. But, in contrast to Kant, I limited the features of intuitive space grasped by pure intuition to certain topological properties; the metrical structure (in Kant's view, the Euclidean structure) and the three-dimensionality I regarded not as purely intuitive, but rather as empirical. Knowledge of physical space I already considered as entirely empirical, in agreement with empiricists like Helmholtz and Schlick. In particular, I discussed the role of non-Euclidean geometry in Einstein's theory."

4.2. General Aspects of the Reception of Einstein's Relativity Among British Philosophers

After having had a glimpse of a few aspects of the early relation of Einstein's relativity with philosophy in Germany, I will move to the specific subject of the present chapter: the reception of Einstein's relativity in Great Britain (basically England). Let me say beforehand, that my discussion is based mainly in the study of a number of books and of two journals, *Mind* and the *Proceedings of the Aristotelian Society*, both recognized as the leading philosophical journals in Britain. It is one of my contentions that this material covers a substantial part of the British philosophical scenario, especially that related to Einstein's relativity contributions, at least in what it concerns the period 1915–1930.

On the basis of such information, several facts emerge. First, that the bulk of the philosophers' reaction concentrated in a very narrow margin, 1920–1922, with not very many contributions before, mainly Alfred North Whitehead's *An Enquire Concerning the Principles of Natural Knowledge* (1919), as well as a couple of articles by Herbert Wildom Carr (1857–1931), professor of Philosophy at King's College, London from 1918 to

1925, one read at the Aristotelian Society in the Session of 1913–1914 (*Proceedings of the Aristotelian Society*, Vol. XIV), and another ("The Metaphysical Implications of the Principle of Relativity") published in the *Philosophical Review* (1915).[15] Of course, what made the difference was the announcement of the results of the 1919 British eclipse expedition, made public at the now legendary November 6, joint meeting of the Royal Society and the Royal Astronomical Society, at Burlington House, London, which made Einstein a world celebrity. Prior to that, very few philosophers paid attention to the special theory of relativity and almost none to the general theory, nor to Einstein's previous attempts to cope with the gravitational field in a relativistic framework.[16] Take, for instance, an article published in 1915 by Charles D. Broad (1887–1971), the philosopher who after having studied at Trinity College, Cambridge, went to St. Andrews University, where he remained until 1920 when he was appointed professor at Bristol University, staying there until 1923, the year he returned to Trinity as a lecturer, a post which gave place to the lecturer in Moral Philosophy at Cambridge Faculty of Philosophy, being appointed finally Knightbridge Professor of Moral Philosophy in 1933, and who participated actively in the philosophical reaction to Einstein's relativity in Britain. The paper, that had the suggestive title of "What do we mean by the question: is our space Euclidean?" (Broad 1915), does not include any mention of Einstein's works on gravitation; actually, Einstein's name is not mentioned at all, even though there is a passing reference to the "modern theory of Relativity in Electrodynamics." The scientists named were Lorentz, FitzGerald and, especially, Minkowski and Robb. It might be of interest to quote the following comment made by Broad: "I do

---

[15] Wildom Carr was for many years Secretary of the Aristotelian Society. He had joined it in the second year of its existence (1881), when he was employed as a broker on the London Stock Exchange. By 1907 he had made enough money to retire and devote himself fulltime to philosophy. As mentioned, he served as professor of Philosophy in the University of London from 1918 (he was then sixty-one years old) until he moved to the University of Southern California."

[16] It must be taken into account in this sense that early in 1915 Adriaan D. Fokker (1915) published an article in *Philosophical Magazine* discussing Einstein and Grossmann's 1913 theory.

142 *Einstein's Relativity in Great Britain*

not say that the facts of electrodynamics do force us to conclude that either Euclidean space or Newtonian bodies are unreal in the present sense; but I take this as an illustration of the sense of reality under discussion, and remark in passing that these facts have actually led certain mathematicians and philosophers, e.g. Minkowski and Mr. Robb of Cambridge,[17] to elaborate a new system of geometry and a new system of physics which shall consistently fit all the facts." That is, Broad, as many others at the time, was particularly attracted by Minkowski's four-dimensional space-time presentation of Einstein's special relativity of relativity.[18] Of course, in general relativity, the four-dimensional structure was ingrained in the theory, but Minkowski was not forgotten. A good example in this sense is what John Synge (1960: ix) wrote in the "Preface" of his textbook *Relativity: The General Theory*:

> "I am much indebted to the well known books by Pauli, Eddington, Tolman, Bergmann, Møller and Lichnerowiz, but the geometrical way of looking at space-time comes directly from Minkowski. He protested against the use of the word 'relativity' to describe a theory based on an 'absolute' (space-time), and, had he lived to see the general theory of relativity, I believe he would have repeated his protest in even stronger terms. However, we need not bother about the name, for the word 'relativity' now means primarily Einstein's theory and only secondarily the obscure philosophy which may have suggested it originally. It is to support Minkowski's way of looking at relativity that I find myself pursuing the hard path of the missionary."

Returning to Broad's 1915 article, we had that it contained assertions which will prove to be incompatible with general relativity, as well as with

---

[17] The work Broad was considering is Robb (1914).

[18] However, it must not be forgotten what Broad said of himself regarding his mathematical knowledge in his influential book *Scientific Thought* (1923a: 4): "In some [...] chapters the reader will find a number of mathematical formulae. He must not be frightened of them, for I can assure him that they involve no algebraical processes more advanced than the simple equations which he learnt to solve at his mother's knee. I myself can make no claims to be a mathematician: the most I can say is that I can generally follow a mathematical argument if I take enough time over it."

The Early Reception of Relativity 143

Einstein's post-1913 metric theories. He said, for instance, that "space and time cannot conceived as capable of causal action on matter."

Broad's paper was commented on the following year, 1916, by John Evan Turner (1875–1947), who taught philosophy at the University of Liverpool until his retirement, and with whom we will encounter later on, but neither in Turner's note nor in Broad's reply — both entitled "The nature and geometry of space" (Turner 1916, Broad 1916) — Einstein's name or the general theory of relativity were mentioned.

A second point that should be stressed is that as far as journals are concerned, it was *Mind* where most articles — the great majority on fact — dealing with the philosophical aspects of relativity — now mainly general relativity — appeared. Based at Oxford and edited by the famous George Edward Moore (1873–1958), this philosophical journal was the organ of the Mind Association, that, at the time we are considering, had 137 members, among them men like Samuel Alexander (1859–1938), the author of the influential *Space, Time and Deity* (1920), F.H. Bradley, H. Wildom Carr, J.S. Mackenzie, J.M.E. McTaggart, Bertrand Russell, as well as A.J. Balfour, the tory politician who became Prime Minister between 1902 and 1905, with Richard Burdon Haldane — also a member of the association and, as we shall see, an important figure in the reception of relativity in Britain — one the great spokesmen of science in the country.

Mind was important for the philosophical discussion of relativity, not only because of the articles it published but also because it gave fast, detailed and informed reviews of many books dealing with Einstein's relativity theories, Charles Broad being, by far, the most prolific reviewer.[19] Thus, among its pages, one finds reviews of Einstein's *Relativity, the Special and the General Theory: A Popular Exposition*, Eddington's *Space, Time and Gravitation*, Erwin Freundlich's *The Foundations of Einstein Theory of Gravitation*, Wildom Carr's *The General Principle of Relativity in Its Philosophical and Historical Aspect*, Cassirer's *Zur*

---

[19] See the bibliography of Broad's writings included in Schilpp, ed. (1959). Besides the Broad's reviews mentioned below, he reviewed another important philosophical book dedicated to relativity that, nevertheless was published a few years later: Émile Meyerson's *La deduction relativiste* (Meyerson 1925, Broad 1925).

144   *Einstein's Relativity in Great Britain*

Einstein'schen Relativitätastheorie, Schlick's *Space and Time in Contemporary Physics*, Haldane's *The Reign of the Relativity*, Ebenezer Cunningham's *Relativity, the Electron Theory, and Gravitation*, Alfred Robb's *The Absolute Relations of Space and Time*, or Hermann Weyl's *Space, Time and Matter*, as well as of Broad's *Scientific Thought*, and Whitehead's *The Principles of Natural Philosophy*, *The Concept of Nature* and *The Principle of Relativity*.[20] It is worthwhile to point out that with the exception of one, all these reviews were published in 1921, with 1920, the pick year, as far as publications were concerned, in the reception of relativity in "philosophical Britain."

A further point to be remarked is that, in fact, it was only a small — though rather active — number of philosophers the ones who contributed, through the pages of *Mind* and the *Proceedings of the Aristotelian Society*, to the philosophical discussion of relativity; the most active of them being Broad, Wildom Carr, Turner, Dorothy Wrinch, and Whitehead, with others like Thomas Greenwood, Ross and Taylor making only occasional appearances.[21]

To the previous list, we must add the names of two physicists: Arthur Eddington and the young F. A. Lindemann, who, as we saw in Chapter 2, together with his father A.F. Lindemann had taken notice of Einstein's new theory of gravitation as early as 1916. However, his role in the philosophical discussion of relativity was almost none.

Of course, it seems more than probable that some of the philosophers "joined the race," to say so, without really being prepared, as in the case of W.D. Ross, mainly remembered today for his magnificent editions of major works by Aristotle, and also, though less, by his works in ethics. In the Oxford Symposium I have just mentioned, Ross confined his comments to the special theory, while the discussion was dedicated mainly to the general theory. He excused his most than probable ignorance of Einstein's new theory of gravitation by saying that "until one can be satisfied about the truth" of the restricted theory "it would be useless to discuss

---

[20] These reviews appeared in Taylor (1921), Broad (1921a), Ross (1921), Broad (1921b), Carr (1921), Broad (1921c), Jeffreys (1923), Wrinch (1924), and Broad (1920, 1921d, 1923b).

[21] There were also cases, like one in which the protagonists were Russell and Strong — on which I will comment later on —, in which Einstein's theory was used only marginally.

The Early Reception of Relativity   145

the general theory which is an extension and in some degree a correction of it."[22] The only argument that really concerned general relativity put forward by Ross was one that had been used several years before — even before the final version of the theory (1915) — by Max Abraham (1913). "Incidentally — Ross stated — one's faith in the argument should surely be somewhat shaken by the fact that the constant relative velocity of light, which is asserted in the special theory, is denied in the general." It was Charles Broad the one who answered Ross's queries.[23]

Another remarkable fact concerning philosophy and general relativity in Britain refers to the temporal distribution of the articles and books published. As noted before, the discussion was initiated in full by 1920 — 1919, if we take into account Whitehead's *An Enquire Concerning the Principles of Natural Philosophy* —, but it reached its climax almost immediately, decaying very rapidly. In the case of articles, the pick year was 1922. Between 5 and 10 papers (depending on the criteria adopted) were published in *Mind*. That year the Aristotelian Society organized an interesting discussion in London, published in the *Proceedings* of the Society the same year (Carr, Nunn, Whitehead and Wrinch 1922), in a volume which also includes two papers, by Whitehead and Greenwood, dealing with relativity. I will analyze that meeting later.

If we focus on the temporal distribution of the publication of monographs, we find it equally concentrated: Whitehead's *An Enquire Concerning the Principles of Natural Philosophy* (1919) and *The Concept of Nature* (1920), Eddington's *Space, Time and Gravitation* (1920), Wildom Carr's *The General Principle of Relativity in Its Philosophical and Historical Aspect* (1920), Haldane's *The Reign of Relativity* (1921), Whitehead's *The Principle of Relativity* (1922), Nunn's *Relativity & Gravitation* (1923), and Broad's *Scientific Thought* (1923).[24]

Several comments can be made on the light of what has been said so far. First, that in some aspects British philosophers reacted to the new theory in the same way as other collectives, although perhaps with more

---

[22] Ross in Eddington, Ross, Broad and Lindemann (1920: 423).

[23] Broad in Eddington, Ross, Broad and Lindemann (1920: 430).

[24] Something similar happened in other countries. In France, for instance, we have that Henri Bergson's *Durée et simultanéité* was published in 1922 (Bergson 1922), though Meyerson's *La deduction relativiste* appeared in 1925 (Meyerson 1925).

146 *Einstein's Relativity in Great Britain*

intensity. I am thinking on the fact that the years 1920–1921 saw the publication of a large number of books — and of general articles, of course — trying to explain the physics of general relativity. That boom affected, therefore, also to British philosophers, or, better, to some British philosophers.

Why, however, that momentum was not sustained?

To answer such a question it, would be necessary to know more about what happened in other countries. However, it can be argued that the loss of momentum suffered by the discussions about the philosophical implications of relativity among British philosophers might have been due to the nature of the topics in which they were interested, and the research programs they cultivated. Theirs was a rather academic philosophy, not particularly dependent in most cases on science. Nothing of the sort of the interests aroused in Central Europe, with, for instance, the Vienna Circle, took place in Great Britain. Perhaps because no Helmholtz, Mach or Boltzmann — that is, scientists with deeply-rooted philosophical interests — had existed in XIXth century Britain, there did not appear men like Schlick, Carnap and Reichenbach. In other words, the philosophy promoted by the Vienna Circle, and, we should add, thinking for example in Reichenbach and Popper, by people subjected to its influence, was particularly well-suited to keep being developed, as it happened in fact, particularly through the methodology of science (the case of Popper), or through the different efforts in the axiomatization of scientific theories.

Related to that situation is the fact that in the 1920s not many of the British philosophers had the scientific training of their German and Austrian counterparts. In general, the British were by education just philosophers; citizens of an academic world dominated by the study of the ancient classics, ethics and moral, philosophy of religion or of politics; it is true that they were also interested in logics and in the foundations of mathematics (Russell and Whitehead), a topic which required a deep mathematical knowledge, but it was a sort of science quite different from physics, in its methodology as well as in its purposes. No one of those that appear in the present study was physicists transformed in philosophers, as was the case with Moritz Schlick and, although to a lesser extent, Hans Reichenbach.

4.3. Einstein and Haldane

Significant in this sense is what Robert Burton Haldane (1856–1928), 1st Viscount Haldane, wrote in his acclaimed book, *The Reign of Relativity* (1921). After pointing out that "Einstein's language is that of the mathematician, and mathematics is his chief instrument," he stated that "into the purely mathematical aspects of such doctrine as that of Einstein, few philosophers are rash enough to attempt to enter. Mathematics talk is an admirably lucid language which is exclusively their own" (Haldane 1921: 39–40).

Haldane's case is an interesting one. He was a British statesman and philosopher. Between 1905 and 1912 he was War Secretary, and subsequently, from 1912 until 1915 when he was forced to resign because of allegations of German sympathies (he had studied, first, at the University of Edinburgh and then at Göttingen University in the 1879s), Lord Chancellor (in 1911 he was created as peer). In Ramsey MacDonald's first short Labour ministry (1924), he was again Lord Chancellor. He was also the first Chancellor of the University of Bristol and elected Lord Rector of the University of Edinburgh, and wrote a number of philosophical books, among them, and apart from *The Reign of Relativity*, *Pathway to Reality* (1903) and *The Philosophy of Humanism* (1922).

Actually, Haldane's book on relativity was his last prominent services to science. Of if, he said: "I projected [it] on the day of my release from office as Lord Chancellor in 1915" (Haldane 1921: xi). A release that was in fact a resignation, brought on by his admiration for the German system of education and science; after August 1914, such admiration marked him as a friend of Germany, and consequently, he was exposed to the hostility and defamation of his fellow countrymen. Fortunately, by 1921 when his book appeared things had changed.

In volume 2 of his biography of Haldane, Frederick Maurice (1938: 97–98) wrote about this endeavor of Haldane:

> "But in 1921 Haldane's chief preoccupation was not political. In May his *Reign of Relativity* was published. Besides the activities which I have mentioned, legal work had been heavy, the Lord Chancellor was abroad at the beginning of the year, and Haldane had frequently to take his place

148   *Einstein's Relativity in Great Britain*

on the Woolsack, he was occupied with a number of important judgements, and he was still constantly travelling up and down the country addressing meetings in education. We may well wonder, with Lord Sankey, how he found the time to complete and see through the press an important philosophical work of 427 pages, dealing with most abstruse problems. It was the most successful book he had ever published. It went through three editions in six weeks, a large edition was required in America, and the book was translated into German, French and Russian. Much of its success was, of course, due to the wide interest which Einstein's statement of his theories had aroused, but the book had its origin not in Einstein, but in the development of Haldane's philosophical thought. It was a natural progression from the second volume of *The Pathway to Reality*. He had begun to meditate on this progression as soon as he left office in 1915 and *The Reign of Relativity* was the result of years of thought and of much reading and research, Einstein had acted as a spur to what was already moving in his mind."

When Einstein visited England in June 1921 — it was his first visit — on his way back of Germany from giving lectures in the United States, he went first to Manchester, where he gave a lecture on June 9, in German and entitled "Relativity," and from there he move to London, where he stayed at Haldane's house, at 28 Queen Anne's Gate (recall the, as a former student in Göttingen, he was fluent in German). Here is how Andrew Robinson (2019: 76–87) explained some of Einstein's activities then:

"Having attended a meeting of the Royal Astronomical Society, at which Eddington gave an account of the solar eclipse expeditions in 1919, the Einstein [he travelled with his wife, Elsa] had dinner at the Haldanes with Eddington, Dyson, Thomson, Whitehead, the Archbishop of Canterbury (Randall Davidson) and others, and met George Bernard Shaw during an after-dinner reception. At some point, Archbishop Davidson said cautiously to Einstein of relativity: 'Lord Haldane tells us that your theory ought to make a great difference to our morals.' But Einstein refused to be drawn (as in his ianchester lecture), and reply simply: 'Do not believe a word of it. It makes no difference. It is purely abstract-science. [...]

The highlight of Einstein's London visit was his lecture at King's College on 13 June in the afternoon — after Einstein, accompanied by

The Early Reception of Relativity   149

Haldane, had laid a wreath on the grave of Newton at Westminster Abbey in the morning."

There, as in Manchester, Einstein spoke in German and about relativity. After the lecture, there was a dinner, during which, Sir Ernest Barker, the principal of King's, said:

"We welcome you twice, for discovering a new truth which has added to the knowledge of the universe, and for coming to us from a country that was lately our enemy to knit again the broken threads of international science. F at your command the straight lines have been vanished from our universe, there is yet one straight line which will always remain — the straight line of right and justice. May both our nations follow this straight line side by side in a parallel movement, which, in spite of Euclid, will yet bring them together in friendship with one another and with a the other nations of the world."

Physics, represented by relativity and Einstein, politics and philosophy, showed themselves poorly differentiated during that first visit of Einstein to England.

Of *The Reign of Relativity*, the least that can be said of it is that it was an extended and well-informed work. In the Preface, Haldane (1921: x) explained its purpose: "The remarkable ideas developed by Einstein, as the result of his investigation of the meaning of physical measurement, have provided fresh material of which philosophy has to take account. [...] The advantage which the methods of science possess is that by them results can be reached and formulated with a precision that is unrivalled, so far as they can go. A price for this advantage has, however, to be paid, and science is apt to find itself in strange regions if it does not limit its scope with genuine self-denial. The inquiry entered on by Einstein has, perhaps because of the presence to his mind of something like this reason, stopped short in his hands of the general problem of the Relativity of all Knowledge. The question that remains is whether the investigation of that problem can be carried further, and if so, whether the philosophical method which appears to be required is a reliable one."

That he was a philosopher of a different sort of, say, a Reichenbach, it is made evident in most of the pages of the book. Take, for example,

150 *Einstein's Relativity in Great Britain*

what he had to say about the principle of relativity (Haldane 1921: 425–426):

> "The real lesson which the principle of relativity of knowledge teaches us is always to remember that there are different orders in which both our knowledge and the reality it seeks have differing forms. These orders we must be careful to distinguish and not to confuse. We must keep ourselves aware that the truth in terms of one order may not necessarily be a sufficient guide in the search for truth in another one. As an aid to our practice, the principle points us in a direction where we may posses our souls with tranquility and courage... The real is there, but it is akin in its nature to our own minds, and it is not terrifying".

Perhaps more clearly, we have what he wrote (Haldane 1929: 345–346) in his autobiography referring to his works, beginning with *The Pathway to Reality* (the Glifford Lectures which were delivered at St. Andrews University in 1903–1904), and later on in *The Reign of Relativity* (1921) and subsequent essays:

> "These works are concerned with what human experience seems ultimately and essentially to imply. They describe an outlook to which, after more than half a century of meditation, I found myself finally compelled. That outlook is not now likely to change substantially.
>
> As against it the physicist may think that he can succeed in resolving the Universe into an assemblage that can be most properly expressed in quantitative equations. For some at least of his critics the question will at once arise as to how the physicist, for whose reflection such an assemblage is present, is himself to be accounted for along with his reflection. For his mind seems to lie at the very foundation if the experience with which he is concerned, actual or possible.
>
> Meditation in this direction led me away from the facile postulates of scientific method in the Victorian period, here and on the Continent. That method of scientific approach does not appear to have been rendered more easy by the recent revolutions in outlook on it which are arising with much insistence in the physics and biology of the twentieth century."

I will say more of Haldane's philosophy later on.

4.4. Einstein's Relativity and British Philosophy

The years — a decade in fact — when relativity was developed were rather special as far as British philosophy is concerned. During the thirty years prior to the advent of Einstein's relativity, the British philosophical scenario had tended increasingly to group itself around two central and directly opposite and contradictory positions. One took "the thinghood of the thing," to use Wildom Carr's no doubt obscure expression as the typical reality, and emphasized the objectivity of existence and the subjectivity of the knowing relations (Carr 1921: 464). The other took the mind and its activity as the immediate intuition of reality, conceiving the fundamental universal reality as the original activity of which the individual mind is the type. The chief influences in consolidating the first, or realistic position (also "new realist" or "neo-realist") came from American philosophers, although there were notable exponents in England — Samuel Alexander, for example, with his book *Space, Time, and Deity* (Alexander 1920) —, while the most striking formulation of modern idealist theory, "or neo-idealist," came from Henri Bergson in France and Benedetto Croce in Italy.[25]

At first, those new realists tried to turn away from the methods of their predecessors, and particularly from those of the idealists; they sought, for instance, to bring philosophy into close relation with science by endeavoring to adopt its modes of investigation. Thus, they gave to the non-mental world the status of being self-subsistent and completely independent of the mind of the observer. "Actual objects did not exist for them in the mind — pointed out Haldane (1921: 261) — but in a medium that is independent of the mind."

Given that situation, it is not surprising that when, in 1919, relativity became a world celebrity a debate arouse among British philosophers — who had felt very crudely the polemic idealism *versus* realism —, as to

---

[25] It might be interesting to recall what Passmore (1966: 301) wrote: "reality, according to Croce, is 'spirit'; to be real, that is, is to play a part in one of mind's diverse activities. Croce opposes any sort of 'transcendence', any suggestion that there is an entity which lies wholly outside the human spirit, whether it be Kant's thing-in-itself, or the Christian 'God', or the naturalist's 'Nature'. Whatever mind cannot find within itself Croce rejects as mythical."

152 *Einstein's Relativity in Great Britain*

which were the philosophical implications, or meanings, of the new physical theories, especially the gravitation theory (general relativity). As we shall see, what characterized that debate is that everybody — realists and idealists, the main contenders, although other positions were also defended — tried to appropriate relativity to favor their own philosophical standpoints. A layman, even a scholar not versed on philosophical matters would have been at lost when reading the different — often philosophically incompatible — statements, counter-statements, and counter-counter-counter-statements made by British philosophers during that brief and intense debate. As the always perceptive Bertrand Russell (1926: 331) noted: "There has been a tendency, not uncommon in the case of a new scientific theory, for every philosopher to interpret the work of Einstein in accordance with his own metaphysical system, and to suggest that the outcome is a great accession of strength to the views which the philosopher in question previously held. This cannot be true in all cases; and it may be hoped it is true in none. It would be disappointing if so fundamental a change as Einstein has introduced involved no philosophical novelty."[26]

4.5. A Stranger in the Philosophical Country: Arthur Eddington

Arthur Eddington, the champion of the relativity, as we saw in a previous chapter, as well as an important contributor to its development, also played a significant role in the philosophical debate concerning relativity in his country.

---

[26] Schlick (1922, 1979c: 344) also noted the same phenomenon when he wrote: "No small number of philosophers have attempted to deny or to minimize this transforming influence; they have adopted a very radical stance towards the theory of relativity. Some, that is, have supposed themselves forced into a flat rejection of Einstein's ideas, as being opposed to their 'commonsense', while others have thought they could derive these ideas from old and long-familiar philosophemes with as little trouble as if they were something quite obvious, and hence constituted no essential advance whatever in philosophical knowledge."

To *Mind* he made two contributions. The first, a paper entitled "The meaning of matter and the laws of nature according to the theory of relativity" (Eddington 1920d), was of no much philosophical interest.[27] With the second, a seminal contribution to a Symposium organized by the Mind Association during the International Congress of Philosophy that was held at Oxford in September 1920, in which, as we already know, also participated Ross, Broad and Lindemann, it was different[28] Actually, both the Symposium in general, and Eddington's article in particular, were a sort of starting point for the debate of the philosophical implications of general relativity in the pages of *Mind*, where the contributions were published.

Eddington's outstanding literary abilities are made evident just quoting his first lines (Eddington, *et al.* 1920: 415), in which he defended the physical dimension of the theory: "It is natural for a scientific man to approach Einstein's theory of Relativity with some suspicion, looking on it as an incongruous mixture of speculative philosophy with legitimate physics. There is no doubt that it was largely suggested by philosophical considerations, and it leads to results hitherto regarded as lying in the domain of philosophy and metaphysics. But the theory is not, as its nature or in its standards, essentially different from other physical theories; it deals with experimental results and theoretical deductions which naturally

---

[27] In the second paper, which I will consider immediately, Eddington summarized this article saying that in it he had tried to show "the exact method by which, starting from a relation undefinable in its absolute character, we arrive from a single source at the physical quantities which describe space and time on the one hand and the quantities which describe things on the other hand. If we describe the character (or geometry) of space and time throughout the world, we at the same time necessarily describe all the thing in the world." Eddington, Ross, Broad and Lindemann (1920: 420).

[28] Eddington, Ross, Broad and Lindemann (1920). Though his participation was not included in the published version, Thomas Greenwood also commented on the relativity subject at the International Congress, as he pointed out in a paper he published afterwards in *Mind*. There he stated that he had "maintained, as against the crude subjectivism of Prof. Eddington and the extreme absolutism of Dr. Ross that the Theory of Relativity cannot be taken as a crucial system to decide between idealism and realism. This view is generally shared by scientists, who cannot profess a great sympathy for the doctrines of those philosophers who endeavour to drive science on to a certain ground which is not its own. I now go further, and hold that the Theory of Relativity is rather a thing prejudicial to idealism, at least as Prof. Carr proposes it" (Greenwood 1920: 205).

154 *Einstein's Relativity in Great Britain*

arise from them." However, he recognized also that there was a domain in it of philosophical interest (Eddington *et al.* 1920: 416): "I would emphasize then that the theory of relativity of time and space is essentially a physical theory, like the atomic theory of matter or the electromagnetic theory of light; and it does not overstep the natural domain of physics. But, speaking to an audience of philosophers, I shall not hesitate to trespass beyond the borderline on my own account. I shall be a stranger in a strange country; and the lurking pits might well intimidate me, if I did not rely on your friendly hands to pick me out."

Actually, Eddington's ideas as put forward in the Symposium and in *Mind* were soon expanded in his classic and very much referred to among British philosophers, *Space, Time and Gravitation* (Eddington 1920c).

It would take us far too long to discuss the philosophical contents of this book, full of comments which could not pass unnoticed to a philosophical mind, as, to put an example, the following one (Eddington 1920c: 197): "Our whole theory has really been a discussion of the most general way in which permanent substance can be built up out of relations; and it is the mind which, by insisting on regarding only the things that are permanent, has actually imposed these laws on an indifferent world. Nature has actually very little to do with matter; she had to provide a basis — points-events; but practically anything would do for that purpose if the relations were of a reasonable degree of complexity. The relativity theory of physics reduces everything to relations; that is to say, it is structure, not material, which counts. The structure cannot be built up without material; but the nature of the material is of no importance."

Haldane was one of the first who reacted to Eddington's ideas as explained in *Mind* and in *Space, Time and Gravitation*, by dedicating several pages of his *The Reign of Relativity* to it (Haldane 1921: 104–121). Again, it would take us to long to enter into this question here, though it must be pointed out that Haldane made efforts to relate some of what Eddington said with, for example, Kant's *Critique of Pure Reason*, as well as with Whitehead's ideas. And let me add, after his "philosophical initiation," that Eddington would made in the future further philosophical comments, not matter that they could be very brief, even in some of his more

The Early Reception of Relativity 155

technical works. Thus, in the Introduction to *The Mathematical Theory of Relativity*, he wrote, in a philosophical vein (Eddington 1923: 3):

> "The study of physical quantities, although they are the results of our own operations (actual or potential), gives us some kind of knowledge of the world-conditions, since the same operations will give different results in different world-conditions. It seems that this indirect knowledge is all that we can ever attain, and that it is only through its influences on such operations that we can represent to ourselves a 'condition of the world.' Any attempt to describe a condition of the world otherwise is either mathematical symbolism or meaningless jargon. To grasp a condition of the world as completely as it is in our power to grasp it, we must have in our minds a symbol which comprehends at the same time its influence on the results of all possible kinds of operations. Or, what comes to the same thing, we must contemplate its measures according to all possible measure-codes — of course, without confusing the different codes. It might well seem impossible to realize son comprehensive an outlook; but we shall find that the mathematical calculus of tensors does represent and deal with world-conditions precisely in this way. A tensor expresses simultaneously the whole group of measure-numbers associated with any world-condition; and machinery is provided for keeping the various codes distinct. For this reason the somewhat difficult tensor calculus is not to be regarded as an evil necessity in this subject, which ought if possible to be replaced by simpler analytical devices; our knowledge of conditions of the external world, as it comes to us through observation and experiment, is precisely of the kind which can be expressed by a tensor and not otherwise. And, just as in arithmetic we can freely with a billion objects without trying to visualize the enormous collection; so the tensor calculus enables us to deal with the world-condition in the totality of its aspects without attempting to picture it."

And he added: "Having regard to this distinction between physical quantities and world-conditions, we shall not define a physical quantity as though it were a feature in the world-picture which had to be sought out. *A physical quantity is defined by a series of operations and calculations of which it is the result*," a phrase that, of course, has an operationalist

156   *Einstein's Relativity in Great Britain*

dimension of the sort Percy Bridgman (1927) made in his famous *The Logic of Modern Physics*, but long before the American physicist wrote it.

4.6. In Defense of Idealism: Wildon Carr and Haldane

I pointed out before that the philosopher's reactions towards Einstein's relativity theory might perhaps depend on the country considered. This becomes clear a propos of idealism, a theme presented intensely in the philosophical discussions dealing with relativity in Britain. Thus, studying the Italian case, Barbara Reeves (1987: 206–208) has explained that the philosophical debate there "must be situated in the context of the neoidealist near-hegemony in philosophy and the longstanding neoidealist devaluation of scientific and mathematical knowledge and research. Epitomized in the *Logic* of Benedetto Croce, published as a book in 1908 but based on a paper first published in 1905, this devaluation denied that science or mathematics were creative activities of the human spirit, that they had philosophical value as knowledge, and they could lead to truth. Scientific and mathematical research resulted not in genuine knowledge but only in classification schemes or techniques useful for practice." And she added the following comparative comment, which is especially interesting for us:

"Therefore, most Italian neoidealists just dismissed the question of the philosophical consequences of relativity, unlike the situation in England, for example. There such idealists as Wildom Carr, Collingwood, Eddington, and Whitehead made relativity considerations central to their writings in this period. Ugo Spirito, young neoidealist follower of Giovanni Gentile, found the efforts of Wildom Carr and his countrymen Alessandro Bonucci and Antonio Aliotta to advocate or at least discuss an idealist interpretation of relativity 'arbitrary' and the idea of an idealist science 'a contradiction in terms,' since an idealist science would no longer be science but philosophy. Furthermore, he denied any connection between the physical principle of relativity and philosophical relativity, finding the physical theory of relativity to be 'only pure idealism'."[29]

---

[29] See Spirito (1921).

The Early Reception of Relativity   157

Once this pointed out, let us go the British case, and in particular to two philosophical books dedicated to relativity that were published in Britain, namely Wildom Carr's *The General Principle of Relativity, in its Philosophical and Historical Aspect* (Carr 1921), and Viscount Haldane's already mentioned *The Reign of Relativity* (Haldane 1921), which shared, in spite of their different contents and tactics, the same philosophical position: A defense of idealism, a philosophical movement of which Bertrand Russell (1917: 588) wrote: "Idealist say all matter is really mind; Materialist say all mind is really matter; American Realists say mind and matter are the same thing, but neither mental nor material." About the New Realism, he said: "In all these respects the New Realism has aimed at inculcating a greater restraint. [...] The New Realism has tried to invent a logical method by which the legitimate conclusions and no more can be extracted from anybody of data. This modest and scientific spirit on the constructive side has perhaps been concealed from readers by a certain arrogance on the critical side, for the New Realism holds that, though there are an infinite number of metaphysical theories which may be true and comparatively few which must be false, it so happens that all the ambitious systems with the exception of that of Leibniz, belong to those few that are demonstrably false."

Haldane (1921: 261), for instance, declared that "the New Realist's point according to which actual objects do not exist in the mind, but in a medium that is independent of mind, is apparently taken to be that of self-subsistent space and time, or of their union in a foundational space-time continuum. For space and time may prove in the end to be only two inseparable forms of a general and self-subsistent externality. Some New Realists go as far as to call space-time the final substance of the phenomenon of experience." On the contrary, he left no doubt that "if the principle of relativity is well-founded the very basis of the New Realism seems to disappear into vapour" (Haldane 1921: 273).

Nevertheless, Haldane's defense of idealism was no so explicit and repetitive as in the case of Wildom Carr, the real champion of that philosophical doctrine in Britain, at least in what it concerns the use of Einstein's relativity theories for its support. In *The General Principle of*

158   *Einstein's Relativity in Great Britain*

Relativity in Its Philosophical and Historical Aspect, Carr (1920: vi) explained how he came to be interested in Einstein's theories:

"My interest in the principle of relativity is purely philosophical, but it is not causal or accidental. I first became acquainted with it at the International Congress of Philosophy at Bologna in 1911, when M. Pierre Langevin, Professor of the Collège de France, revealed its philosophical importance in a remarkable paper entitled "L'evolution de l'espace et du temps."[30] I introduced the subject to the Aristotelian Society in a paper read in the Session of 1913–14 (*Proceedings of the Aristotelian Society*, Vol. XIV), and I contributed an article, "The Metaphysical Implications of the Principle of Relativity," to the *Philosophical Review* of January 1915. Since then the philosophical importance of the principle has received full recognition. It was not, however, until the preparation of my courses of lectures on the 'History of Modern Philosophy' delivered on 1918 and 1919 at King's College, London, led me to read anew the works of Descartes and Leibniz that the quite special historical interest of the main problem impressed me. It is this historical aspect of the principle to which I have tried to give expression in this study. The main idea was developed in a course of lectures on 'Historical Theories of Space, Time and Movement' delivered at King's College in the spring of this year."

It is also worth quoting what Carr (1920: 160–162) had to say in the final paragraphs of his book:

"It seems to me… that the principle of relativity is a philosophical principle which is not only called for by the need of mathematical and physical science for greater precision in the new field of electro-magnetic theory in which it is continually advancing, but is destined to give us a new world-view. It will be found, as it has always been found, that the poets with their mythical interpretations, and the philosophers with their speculative hypothesis, have led the way in this new advance…

I conclude, then, that in every reflection on our actual experience we are directly conscious of an objectivity which we distinguish from our

---

[30] Langevin's conference at the Bologna Congress was published afterwards in *Scientia* (Vol. 10, 1911), and in *La physique depuis vingt ans* (O. Doin, Paris 1923).

subjective activity of knowing. Whether we approach the problem of that objectivity from the abstract standpoint of physical science or from the concrete standpoint of philosophy, the result is the same. Ultimately, in spite of its claim to independence, all that an object or event is, in substance or in form, it derives from the activity of the life or mind for which alone it possesses the meaning which makes it an object or event. This is not a mystical doctrine, nor is it esoteric. If we adopt in mathematics and physics the principle of relativity (and have we any choice?) the obstinate, resistant form of the objectivity of the physical world dissolves to thin air and disappears. Space and time, its rigid framework, sink to shadows. Concrete four-dimensional space-time becomes a system of world-lines, infinitely deformable. And these world-lines, do not they at last bring us in sight of an irreducible minimum of self-subsistent objectivity? No. The world-lines are not things-in-themselves, they are only the expression for what is or may become common to different observers in the relations between their stand-points. Carried to its logical conclusion the principle of relativity leaves us without the image or the concept of a pure objectivity. The ultimate reality of the universe, as philosophy apprehends it, is the activity which is manifested in life and mind, and the objectivity of the universe is not a dead core serving as the substratum of this activity, but the perception-actions of infinite individual creative centres in mutual relation."

4.7. Relativity Debated at the Aristotelian Society

One of the philosophical debates — perhaps the most interesting — in which Einstein's relativity theories were involved took place on 20 February 1922 in the Aristotelian Society, the home ground of the "London Philosophy" as it was called. The debate, presided by Lord Haldane, was dedicated to discuss the possible idealistic (once more) interpretation of Einstein's theories, and had as participants Wildom Carr, the educationalist and philosopher Sir Percy Nunn, Alfred North Whitehead and Dorothy Wrinch. The main purpose of the debate was to discuss a thesis put forward by Carr, which reads as follows[31]: "Einstein's

---

[31] Carr, Nunn, Whitehead and Wrinch (1922). The quotations which follow belong to this reference.

160 *Einstein's Relativity in Great Britain*

theory is a scientific interpretation of experience based upon the principle of relativity. This principle is in complete accord with the neo-idealist doctrine in philosophy, and in complete disaccord with the fundamental standpoint of every form of neo-realism."[32] A thesis in which Carr made no distinction between the special and general theories of relativity.

For Carr, "neo-realism" was the philosophical standpoint according to which "knowledge requires us to presuppose existence, and that in some sense a universe exists in space-time, the entities within which are discoverable by minds, which themselves are accorded a place therein on equal terms with the entities they discover," while "neo-idealism" meant that "reality in its fundamental and universal meaning, is not an abstract thing opposed to nature, or an entity with its place among other entities in space and in time, it is concrete experience in which subject-object, mind-nature, spirit matter, exist in an opposition which is also a necessary relation."

His arguments in favour of neo-idealism were rather simple: "The principle which Einstein follows in physics is based in the recognition that the phenomena which constitute its subject-matter are presented in the form and only in the form of sense-experience. Ultimately and fundamentally the qualities of physical objects are sensations. In this he avows himself the follower of Mach." Classical mechanics, on the contrary, was based in the neo-realist affirmation of an existence independent of sense experience to which the subject of experience referred his sensations. Thus, "the principle of relativity [...] rejects in physics the metaphysical

---

[32] These words were quoted by Dorothy Wrinch, another participant at the Symposium, in her obituary of Wildom Carr. After quoting them, she added (Wrinch 1930–1931: 290): "According, however, to the papers of Sir Percy Nunn, and Professor Whitehead, it was by no means necessary for the realists to modify his position, on account of the principle of relativity. Their arguments, following on that of Professor Carr suggest the inference that the philosopher's attitude to relativity will depend upon the position from which starts." Again, the same philosophical malleability we pointed out before using a text of Russell: "There has been a tendency, not uncommon in the case of a new scientific theory, for every philosopher to interpret the work of Einstein in accordance with his own metaphysical system, and to suggest that the outcome is a great accession of strength to the views which the philosopher in question previously held."

The Early Reception of Relativity 161

principle of materialism which presupposes an objective transcendent cause of experience."

However, Carr did not stop here, with physics; he also thought that the principle of relativity equally rejects in mathematics "the metaphysical principle of intellectualism which presupposes pure reason, enlightenment, discernment, as the transcendent subjective cause of experience. That is to say, it rejects the view that in mathematics the mind, endowed with reason, contemplates eternal truth."

While saying this, Carr was contesting, albeit in an elliptic and indirect manner, Bertrand Russell's main philosophical standpoint, or, in other words, he was denying him the right of using the principle of relativity to support his own views, realism, something that was not accepted by the Cambridge and Bloomsbury philosopher and logician.[33] To see that Carr's statement entered into conflict with Russell's ideas, it is enough to say that he claimed that on the basis of its ability to treat the self-contained character of the world as non-mental and as including universals, he was able to put the connection of logic with mathematics on a new footing. If relations are not merely the products of thought, but confront in the world of experience as existent there not less truly and independently of ourselves that the particulars of sense, then the task of logic must be to investigate these relations. And, again according to him, because those relations are extra-mental entities, notwithstanding their quality of being universals, we can rely on their validity when, by thought and experiment directed by thought we have discovered them, and they may therefore legitimately guide us in forecasting the behavior of the particulars of the experience in which they are embodied.

Coming back to Carr, I will add that as his conclusion he claimed that all he has been saying was "in essentials the Leibnizian conception. The principle of relativity proposed in science precisely the methodological reform which Leibniz proposed in philosophy when he said, 'The monads are the real atoms of nature'."

Carr's arguments were contested on different grounds by the other participants in the discussions, Nunn, Whitehead and Wrinch. Percy

---

[33] According to Haldane (1921: 273), Bertrand Russell was "another brilliant exponent of the doctrine of the New Realism school in philosophy."

162   Einstein's Relativity in Great Britain

Nunn, a realist himself but nevertheless not concerned in his reply "to assert that the principle of relativity is incompatible with idealism; only to deny that it is incompatible with neo-realism," argued that, like Einstein, neo-realists "have taught explicitly that the varying appearances of the 'same thing' to different observers are not diverse mental reactions to an identical material cause, but are correlated sense-data or 'events' belonging to a single historical series."[34] "In other words," he added, the neo-realists "have professed the view which Professor Carr seeks to convey by his explanation, that what makes an object 'common' is to point to point correspondence between the experiences of different observers." In fact, Nunn explained that "before the doctrine of relativity had been risen above their horizon, neo-realists had already gone a long way to meet it, "and that if some were restrained from going farther it was "because they shrank from breaking with the traditional beliefs about space and time which they shared with the idealists of their day," a limitation which, however, he considered Bertrand Russell had avoided in 1914, in his book *Mysticism and Logic*, in which he had "carried neo-realism right into the camp of the relativists" by declaring that "two places of different sorts are associated with every sense-datum, namely the place at which it is and the place *from* which it is perceived." Thus, Carr's refutations were

---

[34] Professor of Education in the University of London, T. Percy Nunn was best known as an educationalist. He wrote little on philosophy, but that little was rather influential. In particular, his contributions to a symposium on "Are secondary qualities independent of perception?" (*Proceedings of the Aristotelian Society* [1909]; see also his book *The Aims and Achievements of Scientific Method* [1907]), was widely studied both in England, where it struck Bertrand Russell's roving fancy, and in the United States. There, Nunn sustained two thesis: (1) that both the primary and the secondary qualities of bodies are really in them, whether they are perceived or not; and (2) that qualities exist as they are perceived. Nunn was not ignorant about Einstein's ideas; in 1923, he published a book on *Relativity and Gravitation*. It gives an idea of its content what he wrote (Nunn 1923: 5) in the preface: "Books upon the Theory of Relativity which are not philosophical in aim generally fall into one of two classes. They are either popular expositions intended for readers who have next to no mathematics, or else serious treatises presupposing in the student a considerable technical equipment. The present work seeks to fill a modest place between the two subgroup." I will say more of this book later on, when dealing with Whitehead's approach to relativity.

The Early Reception of Relativity   163

transformed, according to such line of argumentation, into confirmations; neo-realism was even a precursor of relativity![35]

As for Alfred North Whitehead (of whom we will say more later on), he saw no problem for the realists to associate themselves with the concepts of the new physics: "Why should a realist," he asked, "be committed to an absolute theory of space or of time?" For him, relativity "actually removes a difficulty from the way of the realist. On the absolute theory, bare space and bare time are such very odd existences, half something and half nothing." In fact, he held that "so far as modern relativity has any influence under the problems of realism, it is all to the advantage of such philosophical systems."

Do note in this last sentence, Whitehead *finesse*: He was not a realist, but a constructive metaphysician, with a solid and wide mathematical and physical background, As a matter of fact, he took the opportunity granted to him by the Aristotelian Society debate to remark that the "realist's main difficulty is, however, not removed [by relativity]. Nature is the apparent world; but after all, appearance is essentially appearance for knowledge, and knowledge is a different order of being from mere nature."

The remaining participant in the discussion was Dorothy Wrinch, a fascinating though rather controversial woman who after having been immersed in the study of mathematics and logic, would left her mark in the field of the architecture of proteins, is not as well-known as she should be. Let me, for this reason, beforehand, say something about her.[36]

Dorothy Maud Wrinch (1894–1976) entered Girton College, Cambridge, in 1913, choosing mathematics as her subject. She was the only Girton woman *Wrangler* (highest ranking among those graduating with honours by grades in the final examinations) in the Cambridge 1916

---

[35] Nunn also made some efforts to clarify what he thought were Carr's misunderstandings: "The physics of Einstein takes no more account of the 'subjective' in experience that did the physics of Newton," he pointed out, although it was true that "expositions, especially popular expositions, of the doctrine of relativity make numerous references to the 'observer,' a expository device to keep, before the reader's mind, the 'vital fact' that what distinguishes the older physicists from the modern ones is that the former believed that 'all observations, by whomsoever made, could be referred to a single space-time framework," while the latter "know they cannot."

[36] For more information about Wrinch, see Senechal (2013) and Abir-Am (1987).

164 *Einstein's Relativity in Great Britain*

Mathematical Tripos. However, her interests were not only mathematics but also philosophy (specially, symbolic logic). Thus, in her first year at Cambridge, she attended Bertrand Russell's lectures at Trinity College on "Our knowledge of an external world," and took Part II of the Moral Sciences Tripos in 1917.

During some years, she was close to Russell and his circle, which provided her with both intellectual and social excitement (when in 1916 Russell lost his Trinity College position due to his opposition to serve his country in the World War I, Wrinch was one of the few students — together with men as Jean Nicod and Victor Lenzen — to whom Russell continued privately to teach mathematical logic). In 1924, she married the physicist and mathematician John William Nicholson (1881–1955), who had recently been appointed fellow and director of studies in mathematics and physics at Balliol College, Oxford, a marriage that did not bring her much happiness (it was dissolved in 1938). In Oxford, Wrinch was able to get a position as lecturer (beginning in 1923–1924) at Lady Margaret Hall, an arrangement that prevailed throughout the 1920s and 1930s, working and publishing on subjects such as real and complex variable analysis, Cantorian set theory, transfinite arithmetic, and mathematical physics (especially applications of potential theory in electrostatics, electrodynamics, vibrations, elasticity, aerodynamics and seismology).

One point of her biography during those Cambridge years is worth recalling. It refers to Ludwig Wittgenstein's famous *Tractatus Logico-Philosophicus*. In 1920, Wittgenstein sent the manuscript (in German) of this seminal work to Russell, with the request of helping in its publication. Russell pored over what was a difficult text, and didn't really understand it, but Wrinch said it was profound. And as Russell was departing then for China to lecture there, he eft Wittgenstein's manuscript in Wrinch's hands. When Cambridge University Press rejected its publication, she managed that it appeared (1921) in a German journal, *Annalen der Naturphilosophie*, edited by the chemist Wilhelm Ostwald. Again, she procured that an English edition appeared published by Kegan Paul in 1922.

Beginning in the early 1930s, Wrinch began to change her interests from mathematics and physics to biology (in 1932, she was a founding member of the Biotheoretical Gathering, a group of philosophical and ideologically minded scientists keen on developing theoretical biology).

Indeed, by 1934 she was absorbed with the biological applications of potential theory, and between 1936 and 1939 made her most important contributions, namely the first theory of protein structure, or the cyclod theory (Wrinch 1937). In 1941, she married the American biologist Otto Charles Glaser and settled in the United States, teaching at Smith College, Massachusetts.

In philosophy, and besides her interest on the philosophical status of relativity, which prompted her participation at the 1920 symposium of the Aristotelian Society, in the early 1920s Wrinch wrote papers on the scientific method, especially on inference, inductive logic, and philosophy of probability, often in collaboration with Harold Jeffreys, while in the late 1920s she addressed topics as the philosophical principles of theories like electron theory and quantum theory, continuing her interest on relativity theory, which she considered a "kind of achievement which stands alone in Scientific Theory" (Wrinch 1927: 163), and venturing into "embryonic sciences" (Wrinch 1929–1930) such as physiology, genetics, psychology and sociology.

As to her participation in the Aristotelian Society debate, in some sense, Wrinch did not really entered in the discussion because for her to decide between "realistic" and "idealistic" one would have to recur to the mind, and on this regard she thought "unjustifiable to conclude that the new concepts of relativity allow any deductions whatever to be made as to the nature of mind." For her, relativity was simply a theory which had taken "as its data the particular facts of the external world and arranging and collating them in general propositions — by means of probability inference — interprets its results in terms of the deductions which can be made from certain facts which are unknown, to other particular facts which may or may not be known. Correlation between different facts is the only aim of science."

The nature of Wrinch's remarks can be understood more easily if we take into account some of her other works. Thus, in the 17 February 1921 issue of *Nature*, in an article authored jointly with Jeffreys, she made clear her opposition to Eddington's presentation of Einstein's theory of gravitation, especially as it appeared in his already mentioned book *Space, Time and Gravitation*. What Wrinch and Jeffreys (1921) could not accept was Eddington's view, according to which coordinate systems are arbitrary,

166 *Einstein's Relativity in Great Britain*

having no real physical importance, and always eliminating themselves in any actual physical process. Eddington was charged with the idea that "the only thing that has physical importance is space-time," an idea that, Wrinch and Jeffreys thought, was not the point of view of physics, where the "co-ordinate systems actually chosen are adopted entirely because they give specially simple forms to relations between measured quantities, and thus are not chosen arbitrarily;" moreover, "the properties of space-time never appear in physical laws; thus it is space-time that eliminates itself when the problems are reduced to terms of measurement." In support of their point of view, Wrinch and Jeffreys mentioned Einstein's book *Relativity, the Special and the General Theory*. There, they said, "Einstein's attitude towards 'space' is closer to ours than to Eddington's [because] he appears to regards space, not as a primary entity of Nature, but merely as a conventional construct, composed of the aggregate of all possible values of the three position co-ordinates. In this form the notion may be useful in theoretical work, but we cannot attribute any ultimate physical importance to a thing we have constructed ourselves."

As to Wrinch's own notion of "space", we can refer to one of her papers, which she published in *Mind* (Wrinch 1922: 204):

"to ask 'What is Space' is not significant. We want to investigate what consequences can be deduced from the 'space properties' of terms; we want to establish as many propositions as possible about the further properties which necessarily belong to any term possessing this set of properties. We want to trace the various alternative sets of space properties which are logically possible and to see how far results which are verifiable can be obtained. It will be very advantageous if propositions are discovered which make it possible to apply tests to decide between alternative theories. It will be of great importance, for example, if the further development of Weyl's theory yields some deduction as to the shift of the lines in the spectrum of the sun which would enable us to make a decision between it and other theories which yield other results as to the shift of the lines. But whichever part of the general investigation is being undertaken, not in the case of any of them is it significant to ask, 'What is Space.' It is the properties and not the intrinsic nature of the space which is the subject of investigation."

It is difficult to think of a more concise statement of Wrinch's analytic and logic-based view of science, than this final sentence: "It is the properties and not the intrinsic nature of the space which is the subject of investigation." On the basis of such manifestation, it is not unfair to think that an important element in Wrinch's interest on the theories of relativity is that they were famous theories, important to physicists, mathematicians, philosophers, and the public at large, which could be subjected to her philosophical analytic methods.

4.8. J.E. Turner *versus* Carr and Haldane

The debate regarding the possible idealist or not meaning of relativity did not end, by any means, with the Aristotelian Society discussion. An interesting example in this sense is the paper written by the Liverpool philosopher J.E. Turner (1922) commenting Wildom's Carr and Haldane's books.[37]

For Turner, it was "fundamentally important to recognize that the scientific theory [of relativity] in itself has at bottom very little bearing on any form of the philosophic principle of relativity for] the scientific theory is concerned, and is concerned only, with certain definitely *limited aspects* of space and time — a limitation which is, of course, perfectly legitimate from the scientific standpoint, but which must none the less be carefully borne in mind in any discussion of the theory's philosophical implications. To pass from these 'physical conceptions' to the philosophical aspects of time and space is to 'cross the Rubicon;' and much of the prevailing confusion is due to the transition being undertaken as though it were of no significance, so that what is true of the scientific concept is also regarded as true of time and space within philosophy." Turner thought that Einstein could not be accused of confusing the terms "scientific" and "philosophic," and in this sense he quoted from is book *The Theory of*

---

[37]Turner, the most prolific of Liverpool philosopher of the time, wrote books as *An Examination of William James's Philosophy: A Critical Essay for the General Reader* (1919), *Personality and Reality* (1926), and *The Nature of Deity* (1927). He was, as we see, not particularly dedicated to science.

168   *Einstein's Relativity in Great Britain*

Relativity the statement that "there was nothing specially, certainly nothing intentionally, philosophical about" his theories, adding that neither Carr nor Haldane had adequately recognized "the crucial importance of the distinction" between philosophy and science.

One of the specific points of the idealist conception that Turner could not accept was the role it assigned in relativity to the mind, to the observer's mind, or, in other word, the implications it drew from the role played by observers in relativity. In *The Reign of Relativity*, Haldane (1921: 83) had wrote in this sense that "To understand Einstein's principle as it applies [...] it is necessary to get out of our heads the persistent assumption that when we look out on the universe of space and time we are looking at something which is self-subsistent. For him spatial and temporal relations in that universe depend on the situations and conditions of observers. The character of space and time is therefore purely relative, and so is their *reality*." But that kind of assertion Turner could not accept because it meant to pass "directly from Einstein's physical concepts to time and space in any philosophical sense," something that he, faithful to his distinction between science and philosophy, denied. The fallacy behind the identification of time and space, philosophically considered, with "their physical conceptions," was, according to him, a consequence of the fact that scientific relativity "is not concerned with categories as *categories*; it introduces no new ones and it dispenses with none of our earlier ones; it merely accepts the basal categories of space and time and then renders more precise their application to natural phenomena — corrects our measurements of time and space intervals, our estimates of mass and energy. If, at the outset of experiences, sensations present themselves as we distinguish them, in relations of space and time, then science, for its more special and limited purpose *must* presuppose them; but its increased accuracy in the employment of its fundamental categories is altogether independent of their nature and validity *as* category."

Summing up his arguments, Turner put an end to his paper with a clear and concise statement in which the intrinsic value pf pure philosophy was at its highest:

"All this implies, finally, that what philosophy has to recognize in scientific relativity is simple an increased degree of accuracy due to the greater exactitude of physical concepts; which means, again, that little, if indeed anything, truly metaphysical is in question at all. The established conclusions of the Theory will contribute to the future Philosophy of the universe; but this involves neither a complete revolution in fundamental concepts, nor any substantial advance in the Idealist view of experience and knowledge. 'Change in standpoint,' once more, 'gives no change in the actual."

Wildom Carr (1922) did not loose much time in replying, "amazed, at what seems to me," as he put it, that "their [that is, his fellow philosophers; Turner, of course, in particular] short-sightedness in imagining that philosophy can be indifferent to this stupendous revolution in science." However, his arguments were not new; he only insisted in what he took to be "the special and important work of Einstein in so far as it affects philosophy," in particular in his view that the principle of relativity rejected an absolute which is independent of experience.[38] Neither was Turner's (1922) new reply to Carr especially illuminating. It is obvious, however, that he had not liked Carr's accusation of "short-sightedness," and thus he rapidly pointed out that by no means he could be charged of being indifferent to the new theory (let us stress once more that, as we are seeing, very few significant British philosophers, whatever their philosophical standpoints, were prepared to be accused of being indifferent to Einstein's new theories), fully admitting "both its outstanding scientific value and its philosophic importance. Still, Turner thought that Carr had misread the significance of those theories "as regards the philosophical principle of relativity."

Turner argued that Carr had made an assertion that relativity could not support when he stated that the absolute is not in the object of knowledge, taken in abstraction, not in the external world, it is in the observer

---

[38] He wrote, for instance, that the superiority of Einstein's scheme from the standpoint of philosophy was "that its construction and constitution are inherent in a never transcend the conditions of actual individual experience... The principle of relativity is not the rejection of an absolute and the affirmation of universal relativity. That would be equivalent to the affirmation of universal skepticism. What the principle rejects is an absolute which is independent of experience, and therefore outside knowledge, an absolute which has to be postulated as the condition of knowledge."

170 *Einstein's Relativity in Great Britain*

or subject of knowledge. "The fallacy [of such arguments] becomes obvious — Turner pointed out — [when one notes that] the relativist's system of reference (which is undoubtedly part of the external world) is transformed into the 'object of *knowledge*' and transferred from the external world to the observer. [...] The Question becomes — Is the relativist's reference system a standpoint furnished by the observer? I venture to think that this epistemological problem is as foreign to many physicists as relativity mathematics is to the majority of philosophers. It is indeed a problem which can never be solved on any purely scientific basis such as underlies the theory. The only science which can be appealed to is the science of knowledge. The issue, that is, is epistemological; it cannot therefore be affected by the scientific Theory in any way." His final conclusion was unequivocal: "if any form of subjective idealism has been *already established*, or is *presupposed*, then the Theory amply confirms that philosophy. On the other hand, the Theory itself cannot substantiate it; it is indeed equally consonant with either objective idealism, realism or even materialism; it is, for philosophy, a benevolent neutral."

4.9. Bertrand Russell

It is now time to consider the case of Bertrand Russell (1872–1970), one the British philosophers traditionally associated with Einstein's theories, although it is true that more often than not he was more a general commentator, if not a divulgator, of them than a deep philosophical analyzer.

It is not clear when Russell first undertook a serious study of Einstein's theories of relativity. A remark he made in a paper he published in 1922 indicates that he was aware of the theory of relativity when he wrote *Our Knowledge of the External World* (Russell 1914) in the autumn of 1913, although he did not then appreciate its importance for the topics with which he was concerned in that book. "As I explained in my book on the *External World* (which, however, laid too little stress on relativity), we have to start with a private space-time for each percipient, and generally for each piece of matter," wrote then (Russell 1922, 1988: 132). However, by the spring of 1919 (that is, before the November 1919 announcement of the eclipse expedition) he had studied Einstein's papers on both the special and the general theories. He was staying in the countryside when the measurements of the light bending due to a gravitational field were made during the total eclipse of the Sun on 29 May 1919. His friend, the Cambridge mathematician John E. Littlewood, had arranged with Arthur Eddington, as we know one of the leaders of the expedition, to cable him as soon as a preliminary study of the data indicated the likely outcome, Littlewood then cabled Russell: "Einstein's theory is completely confirmed. The predicted displacement was 1".72 and the observed 1".75 ± 0.06." That cable aroused great excitement in Russell's party.[39] Eight days after the public announcement of the results (6 November 1919), Russell (1919) published his first essay on relativity theory. Many others were to follow.

"In none of these papers, or in his two books, *The ABC of Atoms* and *The ABC of Relativity* — we read in the "Introduction" (Slater 1988: xviii) of one of the volumes that assembled his writings — does Russell claim to be presenting original views of his own. He regards himself merely as an expounder, both to the world at large and to the narrower world of philosophers, of a very difficult set of theories. His intention is to increase awareness of the new physics — philosophers in particular had not, he thought, seen just how important it was in relation to some of their traditional concerns — and at the same time to remind his readers that, because the method of tensors does not lend itself to popular exposition, what they are getting is not the sort of understanding that he enjoys, since he knows the mathematics involved, but only a second best. His younger readers, he hoped, would be motivated to undertake the study of mathematics."

Anyhow, he too used relativity to support some aspects of his philosophical system, So it happened at least when C.A. Strong (1922) objected to his theory of the external world; more concretely to his theory of

---

[39] In his autobiography, Russell (1968: 97) wrote in this sense: "The general theory of relativity was in those days rather new, and Littlewood and I used to discuss it endlessly. We used to debate whether the distance from us to the post-office was or was not the same as the distance from the post-office to us, though on this matter we never reached a conclusion. The eclipse expedition which confirmed Einstein's predictions as to the bending of light occurred during this time, and Littlewood got a telegram from Eddington telling him that the result was what Einstein said it should be."

172 *Einstein's Relativity in Great Britain*

perception, put forward in his books *The External World* and *The Analysis of Mind* (1921), which, according to Russell, amounted to his "theory of physics."[40]

Strong was surprised to find Russell supposing particulars which were members of different pieces of matter to exist all at the same place, in case the place is one reached by light from also different pieces of matter. He also could not understand how objects could be "apparently everywhere except in the place where we see and feel them," and that "a multitude of events, all happening after 12 o'clock, should be the constituents of an event happening at 12 o'clock." To answer to these, and other, queries, that, Russell conceded, "are somewhat curious," but of curiousness due to "modern physics, for which I am not responsible (I wish I were)," he recalled the basic points of the general theory of the gravitational field as well as the Maxwellian synthesis of the electromagnetic field, pointing out that maybe these two fields could be reduced to one "if Weyl is right." Due to the existence of those two fields, "there are — he stated — a number of things happening everywhere always;" moreover, "what we call one element of matter — say an electron — is represented by a certain selection of the things that happen throughout space-time, or at any rate throughout a large region," and "we cannot speak in any accurate sense of 'history' of a piece of matter, because the time-order of events is to a certain arbitrary and dependent upon the reference-body." Such facts were perhaps strange ones, but so was it, and therefore Russell felt obliged to conclude that "it is into a physical world of this description that we have to fit our theory of perception;" in particular, he wished "to include nature and mind into one single system, in a science which will be very like modern physics, though not at all like the materialistic billiard-ball physics of the past." Thus, it seems, that modern physics, in general, and relativity, in particular, played an important role in the configuration of Russell's theory of perception. Also, it must be stressed that his role in the early reception of relativity in philosophical Britain was not as great as his later activity in that field may suggests.

---

[40] Strong was a "critical realist," who in a long series of books, which began in 1903 with *Why the Mind has a Body*, and terminated, after many shifts on points of detail, with *A Creed for Skeptics* (1936), tried to construct a "pan-psychist" ontology.

The Early Reception of Relativity   173

4.10. Alfred North Whitehead's Theory of Relativity

Finally, we get to Alfred North Whitehead (1861–1947), Russell's companion in their masterful though finally frustrated *Principia Mathematica* (1910, 1912, 1913).[41] He was probably the most original contributor to the world of relativity in British philosophy (and perhaps in any philosophical scenario).

What he called his "philosophical relativity" is contained in three books: *An Enquire Concerning the Principles of Natural Knowledge* (1919), *The Concept of Nature* (1920), and *The Principle of Relativity* (1923).[42]

That Whitehead's ideas owed much to Einstein's relativity is obvious just by reading the preface of *An Enquire Concerning the Principles of Natural Knowledge*. There he wrote (Whitehead 1919: v–vi):

"Modern speculative physics with its revolutionary theories concerning the natures of matter and of electricity has made urgent the question, What are the ultimate data of science? [...] The critical studies of the nineteenth century and after have thrown light on the nature of mathematics and in particular on the foundations of geometry. We now know many alternative sets of axioms from which geometry can be deduced by the strictest deductive reasoning. But these investigations concern geometry as an abstract science deduced from hypothetical premises. In this enquiry we are concerned with geometry as a physical science. How is space rooted in experience?

The modern theory of relativity has opened the possibility of a new answer to this question. The successive labours of Larmor, Lorentz, Einstein, and Minkowski have opened a new world of thought as to the relations of space and time to the ultimate data of perceptual knowledge. The present work id largely concerned with providing a physical basis for the more modern views which have thus emerged. The whole

---

[41] For Whitehead's biography, see Lowe (1985, 1990), and for his philosophy of science, Palter (1960). Also Tanaka (1987).

[42] Mention should also be made to an article that Whitehead (1915–1916) published in the *Proceedings of the Aristotelian Society* several years before. It was dedicated mostly to the special theory of relativity, though gravitation was also considered, and Einstein's name made very few appearances.

174 *Einstein's Relativity in Great Britain*

investigation is based on the principle that the scientific concepts of space and time are the first outcome of the simplest generalizations from experience, and that they are not to be looked for at the tail end of a welter of differential equations. This position does not mean that Einstein's recent theory of general relativity is to be rejected. The divergence is purely a question of interpretation. Our time and space measurements may in practice result in elaborate combinations if the primary methods of measurement which are explained in this work."

To Whitehead's credit is that he wrote this book before the announcement of the results of the eclipse expedition (the preface is dated 29 April 1919, and contains the following phrase: "at the date of writing the evidence for some of the consequences of Einstein's theory is ambiguous and even adverse"). That is, he was genuinely interested in some philosophical issues that general relativity made prominent, mainly the nature of space, which, according to him, must be uniform, a point that he stressed in the Preface of *The Concept of Nature* (Whitehead 1920: vii): "Einstein's method of using the theory of tensors is adopted, but the application is worked out on different lines and from different assumptions. Those of his results which have been verified by experience are obtained also by my methods. The divergence chiefly arises from the fact that I do not accept his theory of non-uniform space or his assumption as to the peculiar fundamental character of light-signals. I would not however be misunderstood to be lacking in appreciation of the value of his recent work on general relativity which has the merit of first disclosing the way in which mathematical physics should proceed in the light of the principle of relativity. But in my judgment he has cramped the development of his brilliant mathematical method in the narrow bounds of a very doubtful philosophy."

Whitehead's ideas culminated in *The Principle of Relativity*, in which he put forward a physical and mathematical theory of gravitation, in which he maintained the old division between physics and geometry, not abandoning Minkowski's pseudo-Euclidean four-dimensional space-time (that is, he agreed with Einstein in the belief that the fundamental relations in Nature are not spatial or temporal but spatio-temporal, and that space and time are two abstractions from space-time). We will not enter in the rather well-known fact that Whitehead's gravitational theory led also to the same result of the two first tests of the general theory of relativity, only we will remark that Whitehead continued to use and defend the role of philosophy in the genesis of his formulation. He wrote in this sense (Whitehead 1923: 4–5):

"To expect to reorganize our ideas of Time, Space, and Measurement without some discussion which must be ranked as philosophical is to neglect the teaching of history and the inherent probabilities of the subject. On the other hand no reorganization of these ideas can command confidence unless it supplies science with added power in the analysis of phenomena. [...]

At the same time it is well to understand the limitations to the meaning of 'philosophy' in this connection. It has nothing to do with ethics or theology or the theory of aesthetics. It is solely engaged in determining the most general conceptions which apply to things observed by the senses. Accordingly it is not even metaphysics: it should be called pan-physics. Its task is to formulate those principles of science which are employed equally in every branch of natural sciences."

Certainly, this is what he tried to do, although his approach was a difficult one. Consequently, it is not strange that Whitehead be usually considered an obscure philosopher. Thus, in the obituary he wrote dedicated to Whitehead, C.D. Broad (1948) said that "he was an abominably obscure and careless writer." However, this was not always true and his ideas were carefully considered and appreciated by not a few philosophers, as well as by some physicists. Reviewing his *The Concept of Nature*, A.E. Taylor (1921: 77) stated: "so far as I can judge, Dr Whitehead is fully justified in his contention that his version of [Einstein's] theory is far more consistent and philosophical than any which the physicists *pur sang* have produced."

Although more critical than Taylor, Broad (1923b: 218) could not avoid, in his review of *The Principle of Relativity*, to remark that "What seems to me certain is that Whitehead has produced important arguments which should make us pause before deserting the traditional views so far as to make space-time non-homaloidal. In addition to this he seems to me

176 *Einstein's Relativity in Great Britain*

to have shown quite conclusively that there is nothing to *force* us to a non-homaloidal theory. He has succeeded in giving a modified law of gravitation which will do all that is needed of it, and which requires only the homaloidal Space–Time of the Special Theory of Relativity."

Looking now at the works of other British philosophers, we have that Whitehead's ideas concerning relativity and other topics connected with this subject were often referred to. Those comments oscillated between general and brief sentences like, for example, Bertrand Russell's (1922), "My view of the relation of what we perceive to physics is the same as that of Dr. Whitehead, who first persuaded me to adopt it," to almost exhaustive commentaries such as those Haldane included in his *The Reign of Relativity*, a chapter of which was dedicated to "Relativity in an English form" (it is there where Whitehead's ideas were discussed).

In his book *Relativity & Gravitation*, Nunn, whom we have already met, also commented Whitehead's approach. Indeed, in the preface he pointed out that while writing it he had benefited from the study of Einstein's own papers, "helped out by Professor Eddington's well-known Report, Professor Jean Becquerel's lucid French treatise, and the recently published *Theory of Relativity* of Professor Whitehead" (Nunn 1923: 7).[43] Of Whitehead's book, he added: "From Professor Whitehead's book I have borrowed anything that would fit into my scheme, and I regret that I could not take more. I have compelled to refer to it in the text only very briefly, but have ventured to express the opinion that it is a work of high moment and that its appearance raises issues of critical importance in the mathematico-philosophical discussions which the genius of Einstein and Minkowski set moving."

Later in the book, Nunn (1923: 35–36) had more to say about Whitehead's ideas:

> "We have now carried the description of Einstein's main ideas as far as it
> is profitable to go without the aid of mathematics. But before beginning

---

[43] Jean Becquerel (1922). Becquerel (1978–1953), son of the discoverer of radioactivity and an expert in optics, was professor at the Muséum d'Histoire Naturelle, and also director of the Laboratoire de Physique Générale of the École des Hautes Études.

The Early Reception of Relativity   177

to fill in some details of the sketch we must refer briefly to a theory of relativity different in important respects from the one expounded in these pages. In a triad of very notable books [*The Principles of Natural Knowledge* (1919), *The Concept of Nature* (1920), and *The Principle of Relativity* (1922)] Professor A.N. Whitehead has analyzed the fundamental notions of time, space and matter with unprecedented care and profundity, and, while making full use of the 'magnificent stroke of genius' by which Einstein and Minkowski transformed the old conceptions of space and time, has found himself compelled to take up a critical attitude towards some of Einstein's methods and conclusions. The most salient difference between them is that Whitehead refuses to follow Einstein in attributing physical properties, and therefore heterogeneity, to space. It is a cardinal article of his philosophic faith that temporal and spatial relations must be uniform in character, and that if we assume the contrary we surrender the basis which is essential for the knowledge of nature as a coherent system. But uniformity is not the same thing as uniqueness; there are endlessly numerous time-orders depending on differences in the circumstances of motion of the observer, and there is for each time-order a corresponding space. Logically, time-order is prior to space-order; for space-order is merely the reflection into the space of one time-system of the time-orders of alternative time-systems [*The Principle of Relativity*, p. 8. Alexander (*Space, Time and Deity*, I, pp. 50–8) has much the same idea]. The older physics was right, then, in treating physical phenomena as 'contingencies' superimposed upon the uniformity of time and space. Nevertheless, Einstein is right in contending that laws expressing their character and connexion cannot be true unless they preserve the same mathematical form in all time and space systems."

Nunn recognized that Whitehead's theory "leads to exactly the same predictions, so that experience has produced, so far, no criterion by which the claims of the rival theories may be decided." His book, however, being dedicated to Einstein's theory, Nunn simply added (p. 37) that "there can be no question that Whitehead, in his wonderfully acute and convincing analysis if the fundamental presuppositions of physics — a work that will ever redound to the credit of British thought — and in the theory of relativity he has based on it, has formulated a body of doctrine with which the orthodox relativists must somehow come to terms."

178   *Einstein's Relativity in Great Britain*

Some physicists and mathematicians showed also their knowledge and interest in Whitehead's alternative theory of relativity. George Temple (1926), for instance, wrote in 1926 a paper — "A theory of relativity in which the dynamical manifold can be conformally represented upon the metrical manifold" — based on Whitehead's ideas: "The guiding principles of the theory here exposed appear to be in accordance with the views of Prof. Whitehead, as expressed in his series of three classical works on the principle of relativity." Years later, Temple would return to this theory.

Indeed, *The Principle of Relativity* was also reviewed in the *Philosophical Magazine*, something which indicates that this theory did not pass unadvertised by physicists. The review there began with the sentence (Review Phil. Mag. 1923: 1103): "Professor Whitehead in this volume has put an alternative rendering of the Theory of Relativity which calls for careful and through examination."

Eddington, who had attended Whitehead's lectures when he was a student in Cambridge (Douglas 1956: 10), also showed some sensitivity towards Whitehead's ideas. Thus, in a paper he wrote in 1921 — that is, before the appearance of *The Principle of Relativity* — and referring to the possibility that the curvature of space was everywhere homogeneous and isotropic, Eddington (1921: 803) wrote: "This may appear to have some connection with the view of Dr. Whitehead that space may be Euclidean or non-Euclidean but must be homogeneous throughout." Although, immediately he added: "But uniform and isotropic curvature is by no means a sufficient condition for complete homogeneity of space, and it leaves room for the full range of variation of geometry from point to point required by Einstein's theory."[44] And shortly after the appearance of Whitehead's *The Principle of Relativity*, Eddington published a "Letter to

---

[44] One decade later, in his Presidential Address to the Physical Society, delivered on 6 November 1931, and while commenting on the necessity or not of the cosmological term, λ, Eddington (1932) insisted in his critic to Whitehead's theory: "The ratio of the meter to the radius of curvature is determined by λ. If λ is zero the ratio is zero and the connection breaks down. We are left with a space which does not fulfill the first conditions of a medium of measurement; and the relativity theory is laid open to criticisms such as have been brought forward by Prof. Whitehead (mistakenly, I think, as regards the existing theory) is failing to provide a 'basis of uniformity'." Quoted in Stachel (1986, 2002: 471).

the Editor" in *Nature*, "A comparison of Whitehead's and Einstein's formulae". There, Eddington wrote:

> "In Whitehead's theory of gravitation, as in Einstein's, the tracks of particles in a gravitational field are determined by the condition that certain integral taken along the track is stationary. The integral is denoted by dJ in Whitehead's theory and ds in Einstein's. Light-tracks are further conditioned by dJ or ds, respectively, being zero. Since both theories are known to give the observed results for the perihelion of Mercury and the deflexion of light, dJ cannot be widely different from ds in the field of a single particle (the sun), but I do not think it has hitherto been noticed that dJ is *exactly* equal to ds.
> [...]
> Since then dJ = ds the tracks of planets and of light are the same in Whitehead's theory as in Einstein's. Divergence can only arise in problems involving the exact metrical interpretation of the symbols — *e.g.* the shift of spectral lines. In Einstein's theory the time as measured by a clock is ds, and neither dt_1 nor dt have any fundamental metrical equivalent; in Whitehead's theory dt is preeminently the 'time,' but I must leave to adherents of his theory the elucidation of what this implies."

However, he concluded by pointing out that: "The formula for dJ and ds no longer agree perfectly if more than one attracting particle is considered, because Whitehead calculates the resulting field by simple superposition, whereas Einstein's formula are non-linear."

But, in 1924 no gravitational test involved more than one particle.

In spite of all his possible difficulty or obscurity, Alfred North Whitehead was the most original thinker and philosopher who figures in the history of the early — and not so early — reception of relativity in Great Britain.

I said "not so early," because Whitehead's relativity theory was seriously considered by several physicists as a possible alternative to Einstein's relativity. Prominent among them was John L. Synge, with whom we have encountered in Chapter 2. Besides the papers that he published dealing with Whitehead's relativity theory — for example Synge (1951₂) —, a very interesting document is the 49 pages long typed manuscript of the lectures he delivered in 1951 at the Institute for Fluid

180 *Einstein's Relativity in Great Britain*

Dynamics and Applied Mathematics of the University of Maryland, when he was Visiting Research Professor there. As this work is little known, I will quote extensively from the very interesting "Introduction" to Chapter 1 ("Basic hypothesis") (Synge 1951: 1–4):

> "In 1922 there appeared a book by the late Professor Alfred North Whitehead entitled The Principle of Relativity (Cambridge University Press). His book contains a clearly formulated theory of gravitation and of electromagnetism in a gravitational field, and so invites comparison with Einstein's General Theory of Relativity which had appeared six or seven years earlier.
>
> In attempting such a comparison, one becomes aware of certain psychological factors. The philosophy of science, being on the whole little discussed among active physicists, is naïve in the sense that many things are taken for granted subconsciously. One may believe, for example, that the laws of nature, including the one of which we today know nothing, lie locked as blueprints in a filing cabinet, and that the achievement of the human mind in theoretical physics is less an act of creation than a successful burglary performed on this filing cabinet.
>
> If that view is accepted, then once the real blueprint has been exhibited to the world, any further blueprints for the same structure brought forward at a later date must be dismissed as forgeries. There is no room for two different theories covering the same phenomena; if one is right, then the other must be wrong. That, it may be claimed, is a fair statement of a view widely, if silently, held. It is a view very difficult to dismiss from our minds, trained as they are in an old and little-discussed tradition.
>
> Anyone who attempts to describe Whitehead's theory of gravitation has to face this attitude on the part of physicists, and if he is a physicist himself his own thoughts are not immune from it. He cannot help feeling that he may be swimming against the tide, or (to change the metaphor) exhuming a mummy instead of trying to contribute to the growth of the living body of physics. To bolster his own confidence, he may be tempted to create an artificial enthusiasm as a device of propaganda in order to win a hearing for a theory which has slipped away into a fairly complete oblivion.
>
> There is another difficulty in describing Whitehead's theory. Whitehead was a philosopher first and a mathematical physicist second.

The Early Reception of Relativity 181

How can one who is no philosopher attempt to describe the work of a philosopher? Certainly he cannot, if the work of the philosopher is philosophy. But if the philosophy is only a wrapping for physical theory, then the mathematical physicist can take a savage joy in tearing off this wrapping and showing the hard kernel of physical theory concealed in it. Indeed there can be little doubt that the oblivion in which this work of Whitehead lies is due in no small measure to the effectiveness as insulation of what a physicist can in his ignorance describe only as the jargon of philosophy. The account of Whitehead's theory given in these lectures is emphatically one in which the philosophy is discarded and attention directed to the essential formulae. And this, it may be claimed, is as it ought to be. No student of Newtonian mechanics should be asked to reconstruct it in the form in which it appeared to Newton himself.

The practical physicist who lives among the facts of observation is naturally impatient of theories except in so far as they assist him in understanding those facts. He is entitled to ask, of Einstein's theory of gravitation and of Whitehead's, whether they adequately explain the facts.

In the case of Einstein's theory, the answer might be put like this: For slowing moving bodies and weak gravitational fields, the Einstein equations yield a close approximation to Newtonian gravitation. Since the velocities of celestial bodies are for the most part slow (in comparison with the velocity of light) and the gravitational fields are weak, we are to expect only minute deviations from what is predicted by Newtonian mechanics. Three small deviations in the solar field are predicted and verified by observation within the limits of the errors of observation; these are rotation of the perihelion of Mercury, deviation of a light ray passing close to the sun, and spectral shift toward the red in a gravitational field.

In the case of Whitehead's theory, all the above statements hold. By what appears to be a rather extraordinary coincidence, there is a formal agreement between the two theories in the matter of particle orbits and light rays, and we find ourselves in the strange position of having two theories which both appear adequate on the basis of observation.

Such a situation is rather unusual in physics. The facts of observation are so numerous that it would appear easy to find the crucial observation which would decide in favor of the one theory or the other, or perhaps controvert them both. The difficulty here lies in the close

182 *Einstein's Relativity in Great Britain*

agreement of both theories with Newtonian mechanics, with the result that the critical differences are necessarily very small.

There is however a difference in the character of the two theories. Einstein's theory is based on a set of non-linear partial differential equations involving a somewhat ill-defined term, the energy tensor, and specific applications are exceedingly difficult to work out. The difficulties are not merely those of mathematical manipulations. There are deeper difficulties in the sense that one is hardly convinced that certain problems are clearly formulated; thus in spite of the ingenious manipulations used in handling the n-body problem [Einstein, Infeld and Hoffmann, Annals of Mathematics 41, 455 (1940) and Canadian Journal of Mathematics 1, 209 (1949)], the question remains as to whether these 'bodies' are singularities in the field, and, if so, what is meant by a singularity in a Riemannian space with indefinite line-element.

In respect of clarity, Whitehead's theory has much to recommend it, because it is not a field theory in the technical sense. The problem of n bodies can be unequivocally formulated, and the difficulties of solving it are purely mathematical. Or to take a terrestrial example, it is possible to formulate mathematically the problem of the motion of a fast-moving particle in the presence of heavy fixed masses on the earth's surface, a relativistic refinement of the problem of geodetic observations. In fact Whitehead's theory of gravitation has an applicability which the General Theory of Relativity lacks by virtue of the fact that the latter is a non-linear field theory.

This would appear an opportune occasion to venture an expression of disagreement with what may be called the 'Filing-Cabinet Theory of Scientific Theories' (as discussed earlier) and to suggest that scientific theories might b viewed as statues or models of which there may be several representing one thing, but definitely man-made and subject to rejection, destruction, modification and cannibalisation."

Besides Synge, Alfred Schild (1921–1977) was among those who paid attention to Whitehead's theory. Actually, Schild was related to Synge: an Austrian Jew, Schild fled Nazi Austria in 1939, and after being interned in England he was sent to Canada, where he studied at the University of Alberta with John Synge and Leopold Infeld, Einstein's former collaborator, who supervised Schild's doctoral dissertation (1946). Indeed, in 1949 he signed with Synge a book, *Tensor Calculus* (University

The Early Reception of Relativity   183

of Toronto Press). After spending eleven years at the Carnegie Institute of Technology, Pittsburgh, in 1957 he moved to the University of Texas at Austin, where he established an important center for relativity studies.

Given Schild's relation with Synge, it is not surprising that he became interested in Whitehead's theory of relativity, paying special attention to its possible generalizations. Thus, in 1956 he published a paper entitled "On gravitational theories of Whitehead's type." "In this paper — Schild (1956: 202) wrote — it is shown that there are an infinity of other gravitational theories [besides Einstein`s general relativity] which also agree with all the observational facts known to-day. These new theories are generalizations of Whitehead's theory and include Whitehead's theory as a special case."

Although Whitehead's original 1922 theory predicted the same results than Einstein's general relativity in the two classical tests — as far as the third test, the gravitational shift, its result, though not identical to Einstein's theory were compatible with the observations made at the time —, some efforts were made to find differences between both approaches. In 1954, Gordon L. Clark (1954), of Trinity College, Cambridge, showed that Whitehead's theory predicted a secular acceleration of the classical center of mass of an isolated two-body system.[45] "In a

---

[45] I have not been able to know much about Gordon Clark, besides during several years he was a member of Trinity College, Cambridge, after which he was attached to Bedford College, University of London. He had collaborated with Eddington in a paper on "The problem of n bodies in General Relativity Theory" (Eddington and Clark 1938). In a letter to Eddington of 28 June 1938, H.P. Robertson, professor of Mathematical Physics at the California Institute of Technology, reacted to this paper writing that "I was much interested in the note of yourself and Clark, particularly since it substantiates some conclusions of my own. This agreement is all the more satisfying since my work was based on the equations published recently by Einstein, Infeld and Hoffmann [EIH], whereas yours is based on a modification of de Sitter's procedure." And Eddington replied to Princeton, where Robertson was them, with a letter which offers some information about Clark: "Clark has just got a fellowship to Princeton for a year, so you will doubtless be interested in him. I am very pleased, as he is just at the right stage to profit most by the opportunities. I think very highly of him. It was he who spotted de Sitter's error." Quoted in Douglas (1956: 112). Two days after his letter to Eddington, Robertson wrote to Infeld in not so polite terms: "Dear Sir Arthur and one of his henchmen men [Clark] claim to have discovered and rectified a mistake in de Sitter n-body computations, which when applied to the

184　*Einstein's Relativity in Great Britain*

recent paper — we read in the abstract of this article — J.L. Synge (1952) has shown that the theories of Einstein and Whitehead predict practically the same phenomena in the solar field. He finds that the rotation of perihelion, the deflexion of light rays and the red-shift in the sun's spectrum are the same in both theories. In this paper the motion of the centre of mass of a double star is investigated. It is found that Whitehead's theory predicts a secular acceleration in the direction of the major axis of the orbit towards the periastron of the larger mass. The possibility of a binary star having such an acceleration has already been considered by Levi-Civita (1937), and he has given an example in which it may become detectable in less than a century. On the other hand, it has been shown by Eddington and Clark (1938) that there is no secular acceleration according to Einstein's theory."

Schild continued interested in Whitehead's theory and its possible generalizations. To that end, he dedicated his participation in the XX International Scholl of Physics "Enrico Fermi", organized by the Società Italiana di Fisica at Varenna, on Lake Como with the subject "Evidence for Gravitational Theories", to "Gravitational theories of the Whitehead type and the principle of equivalence." The published version, detailed and pedagogic, dealt was a long review of those theories (Schild 1962a),

---

2-body problem yields exactly your equations — after making some corrections to the trivial mistakes the dunderheads have made. I have written him the sweetest letter congratulating him on his achievement — having no yen to be taken for either a load of manure or a nigger, I altruistically refrained from anything so vulgar as to claim priority for us. I am enclosing of this masterpiece of gentlemanly correspondence;" quoted in Havas (1989: 260). As Robertson said, Eddington and Clark had obtained EIH equations (except for some trivial misprints and mistakes in sign), but they had submitted their paper nine months after Einstein, Infeld and Hoffmann had submitted their. In the previously mentioned paper, Havas (1989: 268) wrote that "Clark was still [after the Second World War] working on the problem [of motion in general relativity], but published in comparatively obscure journals and was ignored." He might be thinking on the *Proceedings of the Royal Society of Edinburgh* (Clark 1954), but he also published several papers in the *Proceedings of the Royal Society of London*, on subjects such "The derivation of mechanics from the law of gravitation in relativity theory" (1941), "The internal and external fields of a particle in a gravitational field" (1949), "The mechanics of continuous matter in the relativity theory" (1949) or "On the gravitational mass of a system of particles" (1949).

The Early Reception of Relativity 185

but a shorter and more to the point statement, centered in the relevance of Einstein's principle of equivalence — that is, that "in a small region of space acceleration id equivalent to the action of a gravitational field" — to distinguish between general relativity and flat space-time theories, appeared in an issue of the philosophically oriented journal, *The Monist*, dedicated to the "Philosophical implications of the new Cosmology." There Schild (1962b: 23–24) wrote:

> "[The principle of equivalence] is the best verified feature of modern gravitational theory; not only the gravitational red shift, but also the accurate experiments of Eötvös and Dicke confirm it. We shall now show in the following that it divides gravitational theories cleanly into two groups, curved space-time theories such as Einstein's and an earlier theory of Nordström's and flat space-time theories such as those of [H.] Poincaré [Rend. Cer Circ. Mat. Di Palermo **21**, 166 (1906)], Whitehead and [G. D.] Birkhoff [Proc. Nat. Acad, Sci. U.S. **29**, 231 (1943); Bol. Soc. Math. Mexicana **1**, 1 (1944)]. We shall give a strong heuristic argument that flat space-time theories cannot satisfy the principle of equivalence, whereas curved space-time theories, suitably formulated, do satisfy it."

That same year, 1962, Schild (1962c) offered a much shorter presentation of his ideas about theories of the Whitehead type. To the purposes of the present book, the main interest of that paper is that it was published in a book, *Recent Developments in General Relativity*, "dedicated to Leopold Infeld in connection with his 60th birthday." That Schild contributed to that volume is expression of his former relation to Infeld in Canada, the same as happened with John Synge and Felix Pirani, also contributors to the book. It is interesting to note that some British physicists who had and were contributing to the development of Einstein's gravitational theory also published articles there, the cases of Herman Bondi ("Relativity and cosmology"), William B. Bonnor ("On Birkhoff's theorem"), Paul Dirac ("Interacting gravitational and spinor fields"), and Dennis Sciama ("On the analogy between charge and spin in general relativity").

THE INTERNATIONAL ASTROPHYSICS SERIES

General Relativity and Cosmology

G. C. McVittie

Chapter 5

A New Generation: George McVittie, "The Uncompromising Empiricist"

In the preceding chapters, I have considered what can be denominated the "first generation" who came into contact with Einstein's relativity, scientists like Larmor, Lodge, Cunningham, Silberstein, Jeans, Dyson, or Eddington, as well as philosophers (Russell, Whitehead, Broad...). But, as it happens in all levels of life, after those scientists came new generations, the first of them, the "second generation", including people — who still had time to know their forerunners — like William McCrea (1904–1999), Gerald J. Whitrow (1912–2000), or William B. Bonnor (1920–2015). Of that "second" generation, I will pay attention only to George Cunliffe McVittie (1904–1988), a physicist whose approach to relativity theory was somewhat different in important aspects to those of his teachers. Of him (and others) John North (1994: 529) wrote: "Richard Chase Tolman (1881–1948) is a good example of a new type of cosmologist. A graduate of Massachusetts Institute of Technology, Pasadena. The author of the first American textbook on (special) relativity, Tolman took a strong interest in the work of Hubble — who in cosmological terms was one of his nearest neighbourgh — and he wrote a brilliant and influential study of ways in which thermodynamics could be introduced into relativist cosmology. Others with similarly broad physical interest were Eddington, McVittie and William McCrea." And a few pages after, he went on (North 1994: 532–533): "The 1930s saw a movement in cosmology of great value to scientific practice generally. It was not for nothing that Eddington's

190　*Einstein's Relativity in Great Britain*

general writings aroused great interest among the philosophers, especially about the nature of theoretical entities. Of course, some of the problems raised came directly from fundamental physics and theories of relativity, but the question of whether or not the spectral redshifts were true Doppler shifts indicative of velocities was seen to depend on what was meant by distance, and as soon as this question was pondered, the entire network of interrelations between observational data and the concepts of cosmological theory was seen to be highly problematic. In no other branch of science was so much care given to the analysis of the concepts employed in it, and here the names of Eddington, E.T. Whittaker, R.C. Tolman, E.A. Milne and G.C. McVittie are among those deserving mention."

More specifically, and remarking one of the outstanding characteristics (which I will comment later on) of McVittie's scientific approach, Stamatia Mavridès (1973: 7) called him "l'empiriste irréductible" (the uncompromising empiricist").[1]

It is precisely because of all this that McVittie is worth our attention; his case can help to understand better the history of general relativity. Moreover, it happens that his professional life and contributions also offer some light in other questions, like the situation of relativity, cosmology and astrophysics both in Britain and in the United States, disciplines to which he contributed and countries in which he lived.

5.1.　Early Education and the Edinburgh Years

George Cunliffe McVittie was born in Smyrna (Turkey), on 5 June 1904, where his father, of Scottish ancestors although born himself in Blackpool, Lancashire, had built a trading company. The family was on holiday in England when Kemal Attaturk drove the Greeks out of Asia Minor, which they had occupied, destroying Smyrna in the middle of September 1922.

---

[1] "Il nous apparaît — wrote Mavridès in the sentence in which the previous characterization appears — que la Cosmologie ná de sel que si on la confronte aux resultants d'observation. Cette discipline ná de sens que fondée sur le faits. 'Lémpiriste irréductible' qi'est McVittie — ainsi que le désignent ses collèges scientifiques — a mis l'accent, à diverses reprises et notamment dans son livre: *Fact and Theory in Cosmology*, sur cette indispensable dualité de la Cosmologie."

In view of the situation, the McVittie's family settled in England, and after an interruption of one year which he spent helping his father, George resumed his studies, entering Edinburgh University.

While still in Turkey, McVittie became interested in relativity. He was what we could call one of the "sons of the 1919 eclipse expedition," in the sense that his interest in relativity first arouse soon after 1919 by reading articles published on the wake of the enormous social popularity that Einstein's relativity theories got immediately after the results of the eclipse expedition were announced. "My father — he wrote in an 'Autobiographical sketch' (McVittie ca. 1975) prepared at the request of the Royal Society of Edinburgh — was the Honorary Secretary of the British Chamber of Commerce in Smyrna and in the spring of 1922 I was employed as Secretary of the Chamber of Commerce. During this period I learnt typing and shorthand but I continued reading mathematics on my own. My father had obtained for me through his book suppliers a book on Einstein's theory of relativity supposedly at a semi-popular level. Though it excited my curiosity, its contents seemed to me to be not only unintelligible but remarkably close to being nonsensical!".[2]

In Edinburgh, McVittie had as teachers scientists like Edmund Whittaker, whom we already met — he was, recall, professor of Mathematics there —, Charles G. Darwin, professor of Natural Philosophy, and Edward T. Copson, Lecturer in Mathematics. Of the three, the only one who did not make any work on general relativity was Darwin: special relativity was an important element of several of his papers, but only dealing with different aspects of quantum physics, as the wave equations of the electron. Although he cannot be considered a "relativist", Copson, already mentioned, was sufficiently interested in the field to publish in 1929 a paper on electrostatics in a gravitational field (Copson 1928). Nothing compared, however, with the many contributions of Whittaker to

---

[2] A copy of these autobiographical notes is deposited at the Niels Bohr Library of the American Institute of Physics: see Guide (1994: 132–133). I am grateful to Alexei Kojevnikov for giving me access to this document. In an interview that David DeVorkin made to McVittie (1978: 4) in March 1978 he was a bit more precise, recalling that he had read "an article on relativity in the periodical *Engineering*", a journal "chiefly for engineers [that] for some reason, my father used to import."

192 *Einstein's Relativity in Great Britain*

Einstein's December 1915 theory, which I reviewed in Chapter 2. Indeed, in the academic year 1927–1928, already a graduate student, McVittie followed one of Whittaker courses on unified theories,[3] an education that served him well when he went to Cambridge to work under Eddington.

In his autobiographical notes, McVittie recalled something of his years as a student in Edinburgh, characterizing some of his teachers in the following manner: "In due course I graduated with First Honours in Mathematics and Natural Philosophy in 1927. During this time, the men teaching influenced me most were (Sir) Edmund T. Whittaker (1873–1956), the professor of pure mathematics, (Sir) Charles G. Darwin (1887–1962), the Tait Professor of Natural Philosophy, and N. Kemp Smith (1872–1958). Professor of Logic and Metaphysics. Whittaker had a highly polished lecturing style and persuaded his audience that every topic was easily comprehensible. A subsequent reading of one's notes showed that this was not so, at least, not until much further work was done. Darwin's lecturing style was untidy but his asides on the nature of applied mathematics — and of applied mathematicians — and his obvious enthusiasm for the subject intrigued me. I often came away from one of his lectures having understood very little but determined to find out what my chaotic notes meant and what it was that aroused such interest in this man. My introduction to relativity theory came through a course that Whittaker gave in 1926/27. To Kemp Smith's discourses on Locke, Hume, Berkeley and Kant I perhaps owe the germ of my attitude to mathematical physics which, many years later, Stamatia Mavridès (1973) was to describe as that of an 'empiriste irréductible' (uncompromising empiricist)."

To DeVorkin McVittie (1978: 12) told something that adds also about his attitude towards mathematics, a point that it is worth to pay attention, in as much as a characteristic of general relativity during many time was that it was "appropriated" by mathematicians. "I wasn't much good at pure mathematics — he recalled to DeVorkin —. I remember we used to call it 'Epsilonology' because the lecturer, and E.T. Copson, used that word too, so it must have come from Oxford, I should think, or it must have been current there. But Epsilonology consisted of doing what I now realize was the proper proof of theorems, with all the logical rigor that was

---

[3] As stated in McVittie (1978, 1987).

needed. That kind of thing always made me impatient. I felt, and still have felt this all my life. I'm quite sure that it is a very good thing, that Bertrand Russell, was it, in the *Principia Mathematica*, wrote two volumes of a thousand pages each, and at the end of the second volume, I believe he finally concluded that one plus one was equal to two. Well, I decided that it was a very good thing for Bertrand Russell to have taken all that trouble, but for goodness sake, I wasn't going to try to understand how he did it!."

After graduating in 1927, McVittie was awarded the Charles Maclaren Mathematical Scholarship (£200 for three years) and the Nicol Foundation (£50 for one year). The second award involved doing some teaching in the Physics Department and therefore he spent the year 1927/28 as a research student at Edinburgh, attending, as already mentioned, Whittaker's postgraduate lectures. Which unified theory version, it was something McVittie could not recall when, many years later, he wrote his autobiographical notes or was interviewed by DeVorkin. Anyhow, Whittaker's lectures gave him "the idea of what unified field theory was, so it was very easy to go over to the Einstein version of 1928 and to Levi-Civita's" (McVittie 1978: 15).

In his 1928 version of unified theory, Einstein (1928) introduced a new geometry, characterized by the property of distant parallelism, or "parallelism at a distance," as others, like McVittie (1929: 1033), called it, expressed in terms of "n-beings," i.e., orthogonal tetrads, while Levi-Civita (1929a, 1929b) modified Einstein's theory by discarding the concept of distant parallelism with respect to 4 orthogonal vectors of reference, using the concept of congruence and introducing the concept of a "world lattice", which was equivalent to a field of tetrads.[4]

It is no surprise that Whittaker taught such a course; as William McCrea (1990, p. 53) wrote: "Whittaker and his pupils in Edinburgh were probably the only workers in Britain who had been interested in the attempts by Weyl and by Eddington to extend general relativity to accommodate electromagnetism into a unified system. Einstein had not much liked these particular attempts — which indeed had not got anywhere much. [..] Almost certainly it was Whittaker who gave McVittie the idea

---

[4]Einstein's paper and use of teleparallelism is studied in Bergia (1993: 292–294) and Vizgin (1994: 234–258), who also discusses Levi-Civita's contribution.

194 *Einstein's Relativity in Great Britain*

of trying to test this new theory by comparing its consequences with an exact particular example of standard Maxwell–Einstein theory."

Whittaker would have given McVittie the idea of what in due course would be his first publications, but it will be not in Edinburgh but in Cambridge and under Arthur Eddington that he would carried out the idea.

That McVittie went to Cambridge was not due to the fact that Eddington had been the first scientist to study unified field theories in Britain, but a consequence of the status — a sort of "scientific imperialism" — that Cambridge had then in British science, a status that implied that even those of Cambridge University former students who had successfully settled in another university wanted that their best students will go to their old *alma mater* to further their scientific careers. "Whittaker — recalled McVittie 1978: 15–16) — was all for sending people he regarded as his bright students to do something at Cambridge, either to take the Tripos or to do research. For instance, W.V.D. Hodge, the geometer, had left for Cambridge a year or two before I went there. Robin Schlapp went as a research student. He never did very much the rest of his life except be a first-class teacher, and run the Department of Theoretical Physics. And there were a number of others. [...] It was taken for granted that if you showed promise as a mathematician outside Cambridge, you would go, in some capacity or other, to Cambridge to finish off, so to speak." This did not apply, McVittie added, to Oxford, "because Oxford regarded themselves as just as good as Cambridge. Copson for instance, don't go from Oxford to Cambridge."

5.2. In Cambridge with Eddington

Before going to Cambridge, to do his Ph.D., McVittie had written to Eddington, who answered on 1 February 1928[5]:

> "Dear Mr. McVittie
>
> I think I have some acquaintance with the work you have been doing, and I shall be glad to act as your supervisor here."

---

[5] "McVittie papers", University of Illinois at Urbana-Champain. When not said otherwise, it means that I am using this source.

However, when in Cambridge, McVittie (who entered Christ's College) found Eddington remote and unapproachable. Here is what he recalled of him (McVittie 1978: 18–19): "I was Eddington's research student, it's true. This meant that perhaps twice a term, I would cycle out on my bicycle to the observatory on Madingley Road, and be shown by the maid into Eddington's study. [...] Eddington would look up from his desk, and I always had the feeling that he was thinking, "Now, who is this young man and why does he come to see me?" But he always was pleasant. We would chat about something."

In Edinburgh, McVittie had enjoyed Whittaker's and Darwin's manners, to the extent that he was not prepared "for Eddington's remoteness and unapproachability" (McVittie ca. 1975). "It is true — McVittie added on that occasion (his autobiographical sketch), that — by 1928 he was entering those mystical realms of thought that were eventually to produce *Fundamental Theory*.[6] He was preoccupied with these matters to the extent that at one point he set me to work on the cosmological problem, forgetting that G. Lemaître, who had worked with him a year or two earlier, had already solved it."

The history of how Eddington did not remember Lemaître's contribution has been told many times, for example, by Godart (1992), Eisenstaedt (1993) and, specially, by Kerszberg (1989: 335–337), who did not forget to mention McVittie's role, a role that McVittie (ca. 1975) himself recalled in his autobiographical sketch when he stated: "In a letter to W. de Sitter posted in Cambridge on 19 March 1930, Eddington writes — misspelling my name — 'A research student McVitie and I had been worrying at the problem and made considerable progress; so it was a blow to us to find it done much more complete by Lemaître'.[7] I well remember the day when

---

[6] McVittie was referring here to Eddington (1953) that, as it is well-known was published after Eddington's death, which happened in 1944. Whittaker was the editor of the book, adding in such capacity a preface and a few notes. E.T. Copson, then at University College, Dundee, in the University of St. Andrews, and George Temple, read the proofs-sheets.

[7] This letter is also quoted by Kerszberg (1989: 336), in a slight more complete form: "A research student McVittie and I had been worrying at the problem and made considerable progress; so it was a blow to us to find it done much more completely by him (a blow softened, as far as I can concerned, by the fact that Lemaître was a student of mine)." What Eddington was working on with McVittie, Kerszberg added, "was the question of whether Einstein's cylindrical world is stable, using two papers by Robertson [1928, 1929] as a basis."

196 *Einstein's Relativity in Great Britain*

Lemaître's letter arrived and Eddington rather shamefacedly showed it to me."

Although not with a thesis dedicated to cosmological models, McVittie was able to write a dissertation on unified field theories (a subject in which he was, as we have seen, well prepared), which was accepted for the Cambridge Ph.D. degree in 1930. Out of his thesis arose McVittie's first papers, in particular, two he published in 1929: "On Einstein's unified field theory" and "On Levi-Civita's modification of Einstein's unified field theory" (McVittie 1929a, b). In the first, he investigated whether an exact solution (which corresponded to an electrostatic field uniform in direction and nearly constant) of Einstein's 1915 general theory of gravitation was also an exact solution of his 1929 unified field theory, finding out that it was not and that the new theory was not equivalent to the old beyond the first order of approximation. In the second paper, he made the same comparison with the theory developed by Levi-Civita, a modification, as it was mentioned, of Einstein's unified theory.

One may think that on the whole the years he spent at Cambridge were not at all favorable to McVittie: Eddington's remoteness, unapproachability and absent mind, and the need to resort, at the end, to Whittaker's ideas and interests, could be taken as good arguments in this sense. However, the Cambridge period was also, as he himself wrote in his autobiographical sketch, "the only time in my life when I received any formal instruction on astronomy. It consisted of Eddington's course in Stellar Structure. Otherwise," he added, "I was self-taught. The books by Cecilia Payne (later Payne-Gaposchkin), Russell, Dugan and Stewart's textbook, the writings of Harlow Shapley and Edwin Hubble were my main sources of information. I also profited greatly by listening to the discussions at the meetings of the Royal Astronomical Society from 1931 onwards."[8]

However, it was not until McVittie (1978: 23) left Cambridge that he began to change his scientific outlook:

"Then there was more unified theory, and more and more theoretical solutions, either of Einstein's original equations, or modified ones, and

---

[8] Most probably, the books McVittie was thinking about are: Eddington (1914), Payne-Gaposchkin (1925) and Russell, Dugan and Stewart (1926–1927). About this last, and influential textbook, see DeVorkin (2000: 224–229).

I began to say to myself: 'There is no way out of this multitude. There is no reason for preferring one rather than another, the way these chaps are going about it'.

And it then occurred to me, slowly, that there is surely a way of getting some order into this confusion, and that is to look at the observational data, and pick out things by that criterion, and not by what seems reasonable or mathematically elegant, or combines relativity and electromagnetism, or as Eddington wanted to do, combines relativity and quantum mechanics. Let's try and pin it down by observations."

This last word, 'observations', is important, because it has been behind many of McVittie's works in cosmology, although not immediately after leaving Cambridge. Indeed, as I already indicated part of McVittie's importance as a relativist lies precisely in this dimension of his work, which allowed that he be called, as we have seen, an "uncompromising empiricist." We shall, however, return to these points later on.

5.3. Further Career in Britain

Coming back to McVittie's career after leaving Cambridge, we have that immediately after getting his Ph.D. degree he went to Leeds University, as Assistant Lecturer in Mathematics, a position that he held until 1933. There he did some works dealing with Lemaître's model of the universe. Together with William McCrea (whom he met in June 1930 in Edinburgh, where he was a lecturer in Whittaker's department), McVittie pointed out that Lemaître's theory did not distinguish between contracting and expanding solutions (McVittie and McCrea 1931). Following Eddington's idea of instability, they investigated the effect a single condensation and showed that it would produce continual contraction, i.e., cause the universe to collapse. Here is how McCrea (1990: 56–57) recalled the origin of this joint work: "we had all recently learned of Hubble's discovery of the expansion of the actual Universe… Somehow we came to know that Eddington had inferred that the Universe had started as an Einstein static universe which had been disturbed and was now expanding for ever more. So when I got home from Scotland I began to wonder whether the formation of condensations in an initially uniform Einstein universe would serve to initiate expansion. I got some results and send them to McVittie for his opinions — which show that he was the one to consult on the subject... McVittie had evidently been looking at the same problem. He replied very quickly showing great interest, but saying that he preferred his own approach — but also suggesting that we should write a joint paper. [...] But we inferred that the formation of condensations would initiate *contraction*! Or rather, the work seemed to show that the formation of a condensation in a uniform Einstein model would make it begin to contract. McVittie himself then published a paper that seemed to show that the formation of more than a single condensation would initiate expansion."

Indeed, a few months later McVittie (1932) was led to an empirically more satisfactory result by considering an initial state with a large number of randomly located condensations. His analysis showed that the Einstein world would turn into an indefinitely expanding Lemaître world and that the formation of condensations thus might be the cause of the initial expansion.[9]

The expanding universe would be in the future one of McVittie's favorite themes, although he considered it connected with several questions, such as spherically symmetric solutions of Einstein's equations, observational data or Milne's kinematical relativity. Precisely because of such plurality of subjects, and the way in which McVittie combined them, the analysis of his works is not easy at all. Contrary to those scientists who make of a specific subject their life program, McVittie pursued quite a large number of problems within the field of general relativity, cosmology and astrophysics. Indeed, although there are permanent traits in his scientific personality (above all, his insistence in relating theoretical entities with observable magnitudes), one of the characteristics of his scientific career was that he was always alert and receptive to new developments. Indeed, one is tempted to say that his critic nature needed of new ideas and theories to develop himself, to flourish as a scientist. It was his sort of "dialectic scientific personality." Precisely because of this is so interesting to study McVittie's works; that is, because such works provide a sort of mirror, a critical mirror, in which to observe what happened to relativity throughout his life.

---

[9] Both papers, McVittie and McCrea (1931) and McVittie (1932), had been commented by Kragh (1996).

A New Generation: George McVittie, "The Uncompromising Empiricist" 199

The examples in this sense are so many that their study would take too much effort and pages. However, in what follows, and while I continue reviewing his life and career, a few of them will be considered. Let us, then, proceed with his career at the point we had left it.

After Leeds, McVittie went to the University of Edinburgh, where he was temporary lecturer in Mathematics during the academic year 1933–1934. To it followed the University of Liverpool (lecturer in Applied Mathematics, 1934–1936), and King's College London (reader in Mathematics, 1936–1948), carrying out duties mainly connected with undergraduate teaching and with little administrative obligations.

5.4. McVittie and Milne's Kinematical Relativity

I said that among the topics McVittie became interested in was Milne's kinematical relativity, which implied the abandonment of some of the fundamental assumptions of general relativity, in particular the principle of covariance. In Chapter 2, I mentioned that Milne's theory aroused much attention since its appearance, and McVittie was no exception, although he tried to impress his own points of views, through a reinterpretation of the theory, a fact that finally would provoke controversies with Milne and his followers.[10]

"During the 1930s — McVittie (ca. 1975) wrote in his autobiographical notes — E. A. Milne's kinematical relativity was vigorously advanced as an alternative to general relativity. I was able to show that the theory of teleparallelism could be used to generalize kinematical relativity." This he did in papers he published in 1935 (McVittie 1935a, 1935b), as well as in his little book *Cosmological Theory* (McVittie 1937), a work which deserves to say a few words before proceeding.

"Cosmological theory — wrote McVittie (1937: 5) in the Preface — is that branch of physics which deals with the structure of matter in its most bulky and massive state, the whole physical universe being regarded as a single system whose broad features are to be investigated. The subject is necessarily highly mathematical, but, in this introductory account,

---

[10] See on this regard, McVittie (1940, 1941).

200 *Einstein's Relativity in Great Britain*

attention has been concentrated on those developments most easily comparable with observation to the exclusion of others of a purely mathematical interest." That is, he wanted a public not limited to scientists.

Differently from other accounts of cosmology, McVittie began *Cosmological Theory* with a chapter on the extra-galactic nebulae, in which the first question to tackle was that of stellar magnitudes and distances: "The first problem connected with the extra-galactic nebulae," he wrote (McVittie 1937: 2), "is the determination of their distance." As there was fundamentally only one method available, viz. the identification in the nebulae of stars of known brightness, or 'luminosity', and a comparison of the apparent with the true luminosities of the stars, McVittie explained what was involved in such method, obtaining the formula

$$\text{Log}_{10}D = 0,2(m - M) + 1$$

where m is the apparent magnitude of the star, M the absolute and D its distance in parsecs. Soon after he deduced another important formula, this one for the distribution of nebulae in space

$$\log_{10}N = 0,51m - 2,758$$

in which N is the number of nebulae over the whole sky of apparent magnitude $\leq m$.

Only once these questions settled, McVittie went into more traditional topics, like tensor calculus and the principles of general relativity (Chapters 2 and 3), which enabled him to discuss the subject, *his subject*, of the expanding universe Chapter 4), both at a theoretical level as well as in its observational dimension (luminosity-distance relations, the N-δ relation, δ expressing the Doppler shift). To remark that he took the opportunity to insist that "the term 'distance' in an expanding universe is ambiguous so long as the method of measurement is not specified. Distance in terms of measurements with rigid scales is not the same thing as luminosity-distance deduced from apparent magnitudes" (McVittie 1937: 68–69). "This dependence of the meaning of distance on the process of measurement — he added — we shall find emphasized on the kinematical theory of the universe" to which he dedicated the final chapter of his book, but not to Milne's own version, but a generalization of Milne's theory to any Riemannian space that he had constructed.

Why, can we ask ourselves, was McVittie attracted by Milne's approach? The answer was put forward quite clearly in that chapter of *Cosmological Theory* (McVittie 1937: 70):

"In the previous chapters the general theory of relativity provided us with a scheme of ideas which had already achieved success as a theory of gravitation. We found that this scheme, which accounted for the small-scale gravitational motions observed in nature, was equally capable of dealing with the structure of the whole universe. But a moment's reflection will convince the reader that the most striking phenomenon exhibited by the universe, the recession of the spiral nebulae, has very little resemblance to gravitational phenomena as exhibited in the motions of planetary systems, double stars, &c. It is therefore legitimate to inquire whether a theory of the universe can be constructed without an *a priori* appeal to a theory of gravitation. The problem which was set by E. A. Milne, and of which he gave one solution, was that of first building up a theory of the whole universe and then, if possible, of deducing from it the necessity of small-scale gravitational motion."

Faithful to his own empiricist principles, McVittie tried to related his version of Milne's theory to observations, producing an expression for the relation between the number of nebulae, N, with Doppler shift, δ, to which he had dedicated, as we have seen, attention previously. However, when comparing such relation and the corresponding one in general relativity, he concluded (McVittie 1937: 95) that "we have here a *too rapid* increase of nebulae with Doppler shift as compared with observation. But it is very satisfactory that the theory does predict an apparent outward increase in the number of nebulae of the type which the observed counts suggest. It will require much greater certainty with regard to the observed nebular count before [the relation between N and δ that he had developed] can be definitely rejected as contrary to observation."

It was different with the problem of the meaning of gravitation in kinematical relativity, of which McVittie (1937: 100), after reviewing the

202   *Einstein's Relativity in Great Britain*

ideas that had been put forward until them, said that "it still awaits a completely satisfactory solution." Nevertheless, he also concluded that "even if a gravitational theory comparable with that of the general relativity is never attained, yet kinematical theory will have served the important purpose of showing that the recession phenomenon is essentially distinct from the gravitational phenomena of the universe."

However, finally, kinematical relativity did not survive. It is interesting in this sense to quote what Otto Heckmann had to say in a review of cosmological theories he prepared for a volume edited, precisely, by McVittie in 1962. Milne," wrote there Heckmann (1962: 436–437), "discovered as early as 1931 that if one adopts a definite continuous group of transformations and demands a world model to be invariant with respect to this group, then the laws governing the motion of the so-called fundamental observers and of free particles are fixed to a large extent. Milne chose the Lorentz group as a starting point. He found new ways of deriving it from a certain set of axioms, it is true. But he considered his main achievement to be a world model which was constructed 'more arithmetico' as he, himself, said and which he claimed to coincide with nature. The basic group was the rigid frame into which everything had to fit, and exactly this rigidity made it impossible to get away from a completely smooth distribution of matter. No clustering, no individual forms, could be described in the theory without inner contradictions to the basic invariance. Milne gave the proof of this impossibility in 1935; but he, nevertheless, later offered a theory of spiral structure based on the time dependence of this constant of gravitation. He offered two time scales and ascribed each of them to definite microscopic and macroscopic phenomena. But he did not see that his two time scales were possible in relativistic cosmology also, and that their separate connection with two definite phenomena was a dream. A very artistic and elaborate mathematical building was erected, but its relation to nature was no here understood. As far as I know, scarcely any additional paper about."

Nevertheless, and looking at a distance, in his autobiographical sketch, McVittie (ca. 1975) conceded that "E.A. Milne exercised a considerable influence on me, particularly through his emphasis on what is nowadays called 'radar distances.' This drew my attention to the problems associated with the notion of distance in general relativity."[11]

5.5. McVittie and the Notion of Distance

We arrive here at one of the notions to which McVittie would in the future paid more attention: the notion of distance (we have found already some evidence of such interest). Although in one way or another such preoccupation was present early in his investigations, specially when he tried to relate observations with theoretical expressions, an aspect of his interests which we have found already, it was specially during the 1950s and after that he insisted on the importance of being extremely precise when talking about distance. Making use once more of his autobiographical sketch, we have (McVittie ca. 1975): "If distance in a model universe could not be defined in an absolute fashion, yet the concept could not simply be ignored. By the middle 1950s… I had proposed that the notion of distance in cosmology could be made precise by defining it with reference to the method to be used for measuring it."

When referring to "the middle of the 1950s," McVittie added a note with a reference to Chapter 8 of his influential book of 1956 (second edition of 1965), *General Relativity and Cosmology* (McVittie 1956, 1965a) There he stated quite clearly that the "problem of distance is a complicated one in cosmology largely because astronomers in their ordinary work are accustomed to using classical theories in which a Newtonian

---

[11] Related to these points, is what Hermann Bondi (1952: 69) wrote in the first edition of his influential book on cosmology: "General relativity bases itself on the concept of the rigid ruler which enters into its fundamental assumption of the metric. As will be seen later this concept leads in cosmology to the mathematically well-defined but physically somewhat nebulous picture of the 'absolute distance' between 'simultaneous events'. This measurement of intergalactic distances with rigid rulers is much further removed from physical practice than the definition of distance adopted by Milne as fundamental for kinematic relativity, which is, at least in principle, capable of being carried put. Milne proposed that an observer, in order to measure the distance of a second observer, should send out a light pulse, and that the distant observer should respond by sending out a similar pulse as soon as he receives the first one (or, alternatively, by reflecting it)."

204   *Einstein's Relativity in Great Britain*

absolute distance in pre-supposed" (McVittie 1965a: 160)., proceeding then to define the notions of "Mathematical distances," "Distance by apparent size," "Luminosity-distance," and which definition of distance must be used so that the term "velocity of recession" of a source be not ambiguous. His book would be not the only occasion in which McVittie studied from a technical point of view the notion of distance in astrophysics and cosmology, but it would take us too far to enter in this point.[12] Instead, it is convenient to recall that he did not limit himself to technical discussions, but that was also a frequent general expositor of the necessity of being careful with the notion of distance in relativity, astrophysics and cosmology, as in the book he published in 1961, *Fact and Theory in Cosmology*, dedicated to "weld together the astronomical observations relevant to cosmology with cosmological theory without entering into detailed mathematical proofs" (McVittie 1961: 9), and in which he concentrated on general relativity, the steady-state theory and, to some extent, kinematical relativity. Indeed, in very few other problems could he better emphasize the necessity that scientists be good empiricists. It is worth to quote in this sense what he wrote in an article adapted from a paper that he presented on 26 December 1957 at the meeting of the American Association for the Advancement of Science held at Indiana (McVittie 1958: 501):

> "What does an astronomer mean when he says the sun and the earth are, on the average, 92,900,000 miles apart? And what bearing can the two theories of relativity have on the matter? It is question of this kind that I shall try to answer in this article, but I must warn you that I have been described by my scientific colleagues as an uncompromising empiricist. I daresay that this is true, but it is also a little strange, for all my training and all my research work have lain in theoretical astronomy and not at all in the extremely difficult and fundamental task of making

---

[12]Thus, in 1959, he described the methods employed by astronomers to determine distances in the galactic and extragalactic domains; in 1965 the luminosity-distance relations for objects of large redshift in various models of the universe, and in 1974 he tabulated the luminosity-distance, the distance by apparent size and the so called U-distance of an object of given redshift (up to $z = 6$) for seven models of the universe (McVittie 1959, 1965b, 1974).

astronomical observations. Perhaps, however, the theoretician does have an advantage over his colleagues who are engaged in observational or experimental work. He can stand slightly to one side and ask himself: What exactly are these men doing, what kind of significance can be attached to the results of their efforts, and in what way are their data really conditioned by theories? To speak in generalities would, I think, be profitless, and it is for this reason that I propose to concentrate on the question of distance in the solar system and to leave the equally intricate and fascinating problem of distance in the universe at large."

5.6. The Cosmological Constant: Einstein and McVittie

There is another general relativity and cosmologic topic of some importance in which McVittie made clear his point of view rather early in his career: the cosmological constant. Indeed, as we shall see immediately, he had an interchange with Einstein with deserves to be remembered.

The starting point of this story was a paper McVittie published in 1933 under the title "The mass-particle in an expanding universe."[13] There, we read (McVittie 1933: 338–339):

"It has been suggested by certain investigators [Einstein (1931), Einstein and de Sitter (1932)] that the cosmological constant, λ, which appears in these formulae is merely a mathematical device and that its value is indifferent from the physical point of view. They propose putting $\lambda = 0$. This cannot be done, however, without introducing difficulties with regard to the expansion." And here he argued that as the values of the density of pressure calculated in the observer's system at an instant cannot be negative, because of the specific characteristics of the equations he had obtained, the observer "must conclude that the expansion is proceeding subject to a retardation. Exactly the same result follows if $\lambda < 0$. Hence, in either case, he must conclude, firstly, that at some time in the past the expansion started instantaneously with a finite velocity; secondly, that there is a 'retarding force' slowing up the expansion which, obviously, cannot be the initial cause that started the latter. No attempt

---

[13] McVittie's model has been summarily studied by Andrzej Krasinski (1990: 118–119).

206 *Einstein's Relativity in Great Britain*

has been made to account for these properties of the expansion, nor is it easy to imagine how they could arise. We are driven to the conclusion that our observer would necessarily take λ to be a positive constant."

Einstein was not happy at all with such conclusion, and in a letter to McVittie written in May 16, 1936, and after mentioning that "your critical research about the situation of the cosmological problems seem to me very interesting," he pointed out to McVittie a few points.[14] The first one that: "In your investigations you are always careful to introduce the cosmological constant Λ [λ as written by McVittie] in the gravitational equations," after which he included as a footnote in his letter the expressive comment: "mea culpa". Such procedure, Einstein went on, "was considered necessary because it was believed that the quasi-stationary character of the space-time metric should be preserved." "However," he went on, "from a formal point of view, the introduction of the Λ term is something absolutely unnatural and odious [*Vom formalen Standpunkte aus ist aber die Einführung des Λ Gliedes eine durchaus unnatürliche und hässliche Sache*], and also seems physically unjustified after the discovery of the expansion motion of matter. In view of this, it would be most desirable that from the very beginning, you do not introduce the Λ term in your researches, that is, that you made it zero. It seems to me that in this way you could obtain results somewhat more secure."

There was at least another occasion in which he corresponded with Einstein: among his papers, there is a letter from Einstein, dated 7 June 1939, in which the creator of relativity, thanking a previous letter from McVittie (which I have not located), reiterated his opposition to introduce the cosmological in cosmology. However, Einstein was not able to convince McVittie, who would remain faithful to the cosmological all his life, or at least most of it. An example in this sense is what he stated (McVittie 1962a: 446) in the summary which closed the volume that he himself edited in 1962, *Problems of Extra-Galactic Research*, and that was the proceedings of a Symposium of the International Astronomical Union that

---

[14]This letter, as well as others I will use, is deposited among the already mentioned "George C. McVittie Papers."

was held at the Santa Barbara campus of the University of California from in 10 to 12 August 1961:

> "I deprecate the identification of the cosmology of general relativity with the special cases in which the cosmological constant Λ is zero and the pressure is zero also. It is only if we make this quite arbitrary preliminary selection that we can agree with Baum that the red-shift apparent-magnitude relation leads to a single conclusion regarding the model universe, or that we can accept most of Minkowski's numerical results. The restriction to these models out of an infinity of possibilities is equivalent to asserting that the problem of the nature of the universe has already been solved, except for the relatively minor detail of selecting one out of a few very similar alternatives. This procedure appears to be in complete contradiction to the assertions of the observers that there are hardly any reliable data from which the nature of the universe can be deduced — to their satisfaction at least!
>
> I agree with Heckmann that any mathematical sound derivation of Einstein's gravitational equations shows that Λ is present. The theory cannot determine the value of Λ, but we can see how it could be found from observation."

And at this point he, wrote the equations that sustained his arguments.

We see that McVittie, as uncompromising empiricist as always, wanted any argument in favor or against the cosmological constant carefully appraised in connection with observations. His approach was very different from that employed by Hermann Bondi, who opened the discussion following McVittie's intervention with the following, rather disdainful, words (McVittie, ed. 1962: 448): "I feel that a discussion of the cosmological constant is unnecessary. While there clearly are arguments in favor of the term, they are not accepted by every student of relativity. Indeed, Einstein himself was very much opposed to the Λ term."

It is not impossible that Bondi's sharp manifestation would be influenced by McVittie's stand as regards a 1948 theory that had been cherished by Bondi, the steady state theory.

The influence that Hermann Bondi, Thomas Gold and Fred Hoyle's 1948 steady-state theory exerted on the general relativity and cosmology scenario during the late 1940s and most of the 1950s has been the subject

208　*Einstein's Relativity in Great Britain*

of several studies, and I will say more also in Chapter 6.[15] Considering that when the theory was formulated, McVittie was an active member of the British general relativity and cosmology community, and that he already had shown his interest in comparing different cosmological theories with observation data, the case of his "relationship" with the new theory seems attractive. And indeed, it is, as we will see.

5.7. The Steady-State Theory

Before proceeding, and in order to fully understand the next section, I must introduce a theory which produced in 1948 by a trio of physicists, Hermann Bondi and Thomas Gold (1948), on one side, and of Fred Hoyle (1948) on the other: more philosophical the former, more physical the latter. The three were then in Cambridge; Bondi (1919–2005) and Gold (1920–2004) were Austrians, who arrived in Cambridge in the thirties escaping from antisemitism in Austria, while Hoyle (1915–2001) was English.

In his influential book, *Cosmology*, Bondi (1952: 140) summed up their version in the following way:

> "The fundamental assumption of the theory [steady-state] is that the universe presents on the large scale an unchanging aspect [this is the so-called 'perfect cosmological principle']. Since the universe must (on thermodynamic grounds) be expanding, new matter must be continually created in order to keep the density constant. As ageing nebulae drift apart, due to the general motion of expansion, new nebulae are formed in the intergalactic spaces by condensation of newly created matter. Nebulae of all ages hence exist with a certain frequency distribution. Astrophysical estimates of the age of our galaxy do not put it into a very rare class of nebulae.
>
> The theory is deductive in the sense that its conclusions are derived from the cosmological principle, but the very powerful formulation of the principle employed dispenses with the need for additional assumptions."

---

[15] See, for instance, Kragh (1993, 1996, 1999).

As to Hoyle (1948: 372), he wrote in his paper:

"Creation of matter was mentioned about twenty years ago by Jeans [*Astronomy and Cosmology* (1928)] who remarked.

'The type of conjecture which presents itself, somewhat insistently, is that the centres of the nebulae (galaxies) are of the nature of singular points, at which matter is poured into our universe from some other and entirely extraneous spatial dimension, so that, to a denizen of our universe, they appear as points at which matter is being continually created.' Subsequently astrophysical developments have, however, shown little support for this particular form of creation.

More recently Dirac [*Nature 129*, 323 (1937)] has pointed out that continuous creation of matter can be related to the wider questions of cosmology. The following work is concerned with this aspect of the matter and arose from a discussion with Mr. T. Gold who remarked that through continuous creation of matter it might be possible to obtain an expanding universe in which the proper density of matter remained constant. This possibility seemed attractive, especially when taken in conjunction with aesthetic objections to the creation of the universe in the remote past. For it is against the spirit of scientific enquiry to regard observable effects as arising from 'causes unknown to science', and this in principle is what creation-in-the-past implies."

As to Hoyle's version, what really distinguished his steady-state theory was the introduction of a scalar field in Einstein's field equations of general relativity. "I chose — Hoyle (1982: 18) said years after — a scalar for my field, which became known subsequently as the C-field, and I constructed the contribution to the action [which led to gravitational equations similar to those of Einstein] from the derivatives of C, with the field taken to satisfy a coordinate-invariant wave-equation. This procedure fixed the situation uniquely within the framework of classical physics."

Though the steady state has long been discarded once Penzias and Wilson (1965) discovered the fossil remains of the Big Bang, that is, the cosmic background microwave And during the late 1940s, the 1950s, and even in the early 1960s, although by then its influence had decayed, especially outside Great Britain. During such important occasion as the meeting held in Bern, 11–16 July 1955, to celebrate the Jubilee of Relativity Theory (it become considered the "zero" International Conference on

210 *Einstein's Relativity in Great Britain*

General Relativity, that is "GR0", while, as already mentioned, the one held in 1957 at Chapel Hill was "GR1"), Herman Bondi (1956: 152–154) defended the steady-state theory with the following arguments:

> "In the first instance I should like to draw attention to how scientific cosmology has become. The cosmological papers today have all dealt with empirical tests of cosmological theories and nobody has referred to how satisfying or how beautiful or how logical this or that theory is. A few years ago, the emphasis would probably have been the other war round.
>
> My second point follows on from this and concerns the criteria of what a theory should do. The absence of a complete and complex mathematical apparatus in the steady-state theory has been deplored. This attitude seems to me to be incorrect. The only legitimate demand to be made is that a theory should enable predictions to be made that can be checked observationally. If such predictions can be made without cumbersome mathematical apparatus, then this is at least as good as if a mathematical theory is required."

After that general introduction, Bondi referred to four tests which seemed to him "to be most significant and more likely to lead to decisive results:" (i) "the Stebbins-Whitford effect (and its generalizations)," in which the properties of distant and near galaxies are compared; (ii) "the ages of galaxies;" (iii) "the origin of the heavy elements;" and (iv) "the condensation of galaxies". And his conclusion was that "the steady-state theory leads to more fruitful theoretical problems since it has to be shown that present conditions are self-perpetuating, whereas in evolutionary theories an explanation of present features almost invariably involves the postulating of special initial conditions."

However, such arguments did not have a long life.

Before 1965, it was the work of Martin Ryle (1918–1984), lecturer in Physics and Astronomy at Cambridge University since 1948 and professor of Radio Astronomy since 1959, a position which he shared with the direction of the Mullard Radio Astronomy Observatory, that undermined the theory, much to the opposition of Hoyle, who never abandoned his theory.[16] But the first announcements of the refutation of the theory, based on Martin Ryle's radio stars counts, with its "log N-log I curves" (N = number of radio sources per unit solid angle having greater intensity than I), on the basis of which he saw "no way in which the observations can be explained in terms of a Steady-State theory" (Ryle 1955: 137) were received with caution. Speaking at an important conference, held at the University of North Carolina, Chapel Hill, in January 1957, Arthur Edward Lilley (1957), one of the pioneers of radio astronomy in the United States who was working then at the Yale University Observatory, could say: "[Ryle and Scheuer's observations] will suggest departures from an isotropic and uniform universe and the results, if valid, are not consistent with a steady-state universe. However, the interpretation of this [logN-logI] curve has been discussed by Bolton, who has suggested that when one has observational errors which increase with decreasing intensity, even an isotropic distribution can produce a curve of the form [given by Ryle and Scheuer)."

Six years later, the conflict had not yet been settled and William McCrea (1963) could still point out, during his Presidential Address at the Royal Astronomical Society that "M. Ryle and his colleagues showed almost beyond doubt that their radio surveys reach much further into the universe than any present optical surveys can do. The first of Ryle's surveys conflicted greatly with the steady-state mode! His recent surveys conflict less violently but still very definitely. Recent optical surveys appear on the whole to conflict but the conflict is probably not yet outside the limits of uncertainty in the observations. Indeed, the general picture is that the observational results that have from time to time been in gross conflict with the steady-state model have mostly had to be revised and the revision usually seems to reduce the conflict; for the rest the situation taken as a whole is inconclusive."

---

[16] See, for instance, Hoyle (1980) and Hoyle and Narlikar (1974), as well as Hoyle's autobiography (Hoyle 1994). Also Kragh (1996). For more on the sequels of the steady-state cosmology in Great Britain, see Sánchez-Ron (1990).

212 *Einstein's Relativity in Great Britain*

Just two years after McCrea's assertions, Penzias and Wilson added to Ryle dismissal of the steady-state cosmology.

5.8. McVittie and the Steady-State Theory

According to his own recollections, McVittie (1978) learnt about the steady-state theory directly from Bondi: "I was at King's College [remember that he was reader there until 1948] and Bondi came to see me, before he and T. Gold went to a meeting of the Royal Astronomical Society which was held in Edinburgh, as far as I remember.[17] He came and told me about this theory, and I said, 'Well, yes, Hermann, do it that way, but this is much more restrictive than general relativity.' I didn't show any great enthusiasm. However, I was taken aback when, after the meeting of the Astronomical Society where this was first expounded publicly by Bondi and Gold, E.T. Whittaker wrote to me and said that he'd heard the most interesting account, from two youngish men called Bondi and Gold, about a new theory of the universe and so on. So I said to myself, 'ell, dear me, have I missed something?'"

I have been able to locate Whittaker's letter to McVittie. It is dated "2 Nov 1948." Here is what it says:

> "We had a good meeting of the R.A.S. in Edinburgh. Cosmology was prominent, as besides a paper by E.A. Milne (showing how his theory of special nebulae could be extended so as to account for star-streaming) ten was a most suggestive paper, rich in original ideas, by two Austrian Jews who are now Fellows of Trinity [Cambridge], Bondi and Gold. Although I enjoyed and appreciated both Milne and Bondi-Gold, they didn't convince me. I still think that the foundations of the true cosmology were well and truly laid by Eddington and that his 'cosmical number' N is the one discovery so far that is both likely to be permanent."

So, Whittalker was not as enthusiastic as McVittie recalled 30 years later. Anyhow, when the steady-state theory appeared in print, McVittie (1978) "looked at it, and the more I looked at it, the less I liked it. For one

---

[17] Such meeting took place on October 29–30, 1948. It was, according to a official history of the Royal Astronomical Society (Sadler 1987: 119), "a resounding success." This reference contains a photograph of the group of scientists who attended the meeting.

thing, it was very restrictive, compared to general relativity, and for another, it contained this most mysterious creation process."

Indeed, the new cosmological theory, which won lots of adepts in Britain (especially there) during the 1950s — until, as I mentioned, Martin Ryle's radioastronomical counts began to turn the tide — contributed to McVittie's emigration to the United States. "In fact — he recalled (McVittie 1978) decades later — one of the reasons why I went to the United States, I think, was to get away from the atmosphere of the steady-state theory [in Britain]. There was such a hullabaloo about the new revelation! Everybody."

In the years to come, McVittie would remain a firm antagonist to the steady-state theory. Leaving aside his publications, there is ample evidence of such antagonism among his papers. A few examples deserve to be mentioned, although they refer to the period when he was already settled in America, period I will consider later on.

The first example concerns Martin Ryle. On 6th January 1957, McVittie wrote to Ryle:

> "I am sending you by Air Mail parcel post a copy of a paper which I have written on the theory of the distribution in space of extra-galactic radio-sources and which may interest you. I have been working on this on and off since the I.A.U. meeting and was stimulated to finish the work by the receipt of the Cambridge and Sydney catalogues. I have sent the top copy of the MS to [Joseph L.] Pawsey.
>
> Do not assume that, because I have used the Australian data for illustrative purposes, I thereby commit myself to accepting them in preference to the Cambridge ones! On this controversial question, I do not have the necessary technical knowledge to take sides. My reason for selecting Mills and Slee's counts rather than yours is that the former are more easily dealt with by means of what I have called 'first order $\log N$ models" than are the latter. You will see from p. 19 of the MS that, to get the Ryle and Scheuer slope for the $\log N$ — $\log S$ curve, would need so small a negative value of b_1 that second-order models, at least, would have to be used. I feel that the labour involved in using them would not be justified until the Sydney-Cambridge controversy is settled."

And he finished his letter with the following sarcastic words" "Perhaps I must apologize for using general relativity model universe in

214 *Einstein's Relativity in Great Britain*

preference to the Bondi and Gold creation of matter theory, which, as I see it from this distance, is now regarded as almost the final word on cosmology in Britain. Apart from the question of creation, the Bondi and Gold theory so restricts R(t) and k that it is always running up against contradictions with observations."

The second example is another letter, this time one McVittie wrote to Joseph Pawsey, the scientific leader of the Sydney radio astronomers, on 16 April 1958:

"Dear Joe,

I have had some correspondence with Fred Haddock, after he wrote me saying that I was not to [be] a speaker on the program of the Radio-Astronomy Symposium. I gather that there is to be one formal speaker only on cosmological matters and that he is going to be Hoyle. It seems to me that this tantamount to giving the approval of the Symposium to the 'creation of matter' point of view in cosmology and only permitting others views to be expressed in any discussion-period there may be and provided opponents of the 'creation of matter' theory can get a word in. I believe that a similar policy was adopted at the Manchester Symposium.

Let me say at once that I am not suggesting a modification of the Committee's decision, which would be humiliating for all concerned. I would like your views on a personal question that the decision brings up. Perhaps rashly, I took your remark whilst we were motoring to Yerkes last October that you wanted me to read a paper at the Symposium, at its face value and have given a good deal of thought to the problem you spoke about last year, namely, how can one deal with a scatter in the intrinsic flux-density of extra-galactic radio-sources? I have got to the stage of modifying the theory I put forward in the Austr. J. of Physics paper,[18] to allow for an arbitrary mixture of standard sources. Essentially this is to be done by a step-function distribution of flux-densities, to replace the single standard flux-density used previously. The work still needs a good deal of polishing up, which I had intended to do in May and June. But Fred's letter contains the remark about the Symposium that 'a small amount of time should allocated to cosmology because of the very uncertain and tentative nature of the

---

[18] He must refer to McVittie (1957), where he made use of both the Cambridge and the Sydney catalogues of radio sources.

observational radio data upon which to base cosmological conclusions.'
I suppose that this is a hint that the kind of investigations I have carried
out — or in the course of doing — which are intended to show how radio
data could be useful, are not of interest to radio-astronomers, but that
'creation of matter' speculations are. If this is the correct interpretation
of the view of radio-astronomers, I would drop the work I have been
doing like a hot brick and turn to something else. But I wanted to check
with you first that this was so."

The Radio-Astronomy Symposium to which McVittie was complaining was held in Paris in August (1958), and we have a copy of a report to the Office of Naval Research, that financed his attendance to it, as well as to the Xth General Assembly of the International Astronomical Union, held the same month in Moscow, which deserves to be quoted, as it provides an interesting personal, but informed, perspective of events that were happening a the time in the field of gravitation, astrophysics and cosmology.[19]

After complaining about the accommodation facilities provided by the Paris Symposium, McVittie summarized his opinion about the scientific content of the meeting:

"The Symposium revealed that the leadership in radio-astronomy is still
held by the groups at Leiden (Holland), Cambridge and Manchester
(England) and Sydney (Australia). In my own fields of interest —
galactic and extragalactic studies — the most remarkable piece of work
carried out in the U.S. was by D.W. Dewhirst during a visit to the
Mount Wilson and Palomar Observatories from Cambridge, England.
His identifications of peculiar galaxies with faint radio-sources is a
considerable step forward in this difficult subject. A feature of the
groups from Leiden, Cambridge, Manchester and Sydney was the existence of teams of young workers in each place. The proximity of a
University to each of these centers of radio-astronomy is significant.
The University can provide the training in physics, mathematics,

---

[19] G.C. McVittie, "Report on attendance at Radio-Astronomy Symposium, Paris, and Xth General Assembly International Astronomical Union, Moscow, August 1958." "McVittie papers. Urbana."

216 *Einstein's Relativity in Great Britain*

astronomy and electrical engineering needed by young men who can then become radio-astronomers.

I read a paper entitled 'Remarks on Cosmology' at the Symposium, which will be published in its Proceedings.

The development of radio-astronomy in France is startling. The equipment at Nancy is lavish and the men are keen and able. The same may be said of optical astronomy at the Observatoire de Haute Provence, where I spent three days. There the new reflector of approximately 72 inches had just come into use and excellent trial photographs had been obtained. The optical image-amplifier due to Lallemand and Duchesne appears to be a workable and very valuable device. At the Observatoire de Haute Provence Laffineur has constructed a radio interferometer using two parabolic cylinder antennae. It has a base-line of 1 km. The instrument was crudely constructed of chicken-wire on wooden posts."

Of the Moscow meeting, McVittie, who acted as Secretary of one the Commissions (No. 28, dedicated to "Extragalactic Nebulae"), selected to say that "the most remarkable work reported on at the Commission 28 meetings was by a Russian, A.L. Zelmanov, who has been working on non-homogeneous models of the universe."[20]

More informal and direct is what McVittie told he great Allan Sandage in a letter of 23 September 1958, which will serve as my final example, and in which, together with other questions, McVittie's antagonism to the "steady-state boys" is evident:

"I had a notice the other day of the Neighbors meeting at Perkins on Oct. 4 together with a covering note from Sletteback urging me to come because 'Allan Sandage especially mentioned you in his letter.' I like to think that this really happened — even if it did not — because I have been feeling recently that what I might have to say about cosmology is not of interest to the younger generation! Perhaps it is because I am still feeling a bit low as a result of a parting gift from our Russian friends in Moscow which was a mild bout of pneumonia from which I recovered in England. In any case the summary of your proposed talk interested me greatly and I should much have liked to be able to come on Oct. 4. But

---

[20] It must be recalled that he was unable to attend the meetings held during the last three days, because he contracted a mild case of pneumonia.

I am supposed to take things a bit easily for a time and I have the long journey to Green Bank to face on Oct. 14–16.

In the summary of your lecture there is a reference to a 'theory of gravitation' which could be used, in part, to find the 'mass of the universe'. I was intrigued to know what this theory could be: is it a new one? Also you seem to imply that the mass of the universe is finite and therefore that the curvature is positive. In the general relativity interpretation I long ago concluded that today's data, whether on red-shift or by counts of faint radio-sources, cannot determine the sign of the curvature directly as it is a third-order effect. Of course, one can help oneself out with Einstein's equations but then the balance of evidence seems to be in favor of negative curvature.

The status of the deceleration also puzzles me. The steady-state boys (Burbidge, Gold, Hoyle) at the Radio-Astron. Paris symposium noisily and rather rudely insisted that the deceleration no longer existed. Mayall and I were both puzzled by this hullabaloo. [...] Also in Moscow, Baum gave us a talk on this large red-shifts and drew the conclusion that the deceleration occurred. So what were the steady-state boys shouting about, and on whose authority?

Apparently the new party line us that faint radio-sources are not extra-galactic, at least so Gold told us making little effort to conceal his irritation. This after arguments by Mills, Ryle and Dewhirst to show that they either all were or that at least some of them were (Dewhirst). I think I am right in saying that Dewhirst's work impressed greatly most of the participants in the symposium. So again I am puzzled by what may cooking!"

5.9. General Relativity and Cosmology

In 1948, McVittie was appointed professor of Mathematics and Head of the Mathematics Department at Queen Mary College, a position that offered him a little more scope as regards advanced teaching work that what he had enjoyed before.[21] There he had Clive W. Kilmister (1924–

---

[21] The fact that McVittie was attached to a Mathematics Department must be understood on the light that during many years it was frequent that relativists were members of such departments (a notable example is the Department of Mathematics of King's College London, in which worked during the 1950s, 1960s and 1970s men like Herman Bondi and Felix A.E. Pirani.

218 *Einstein's Relativity in Great Britain*

2010) as his first Ph.D. student, who in due course, years afterwards, would become professor at the Mathematics Department at King's College London, and a frequent commentator of Eddington's later works.

It was during his years at Queen, almost at the end of they, that began the process that would lead to the publication of one of the most well-known and influential works: the book *General Relativity and Cosmology*. The origins of this work can be dated on 29 March 1950, when M.A. Ellison, from the Edinburgh Royal Observatory wrote McVittie telling him that "Lowell and I are co-operating as Editors, with Messrs. Chapman & Hall in the publication of a series of monographs, under the general title 'International Astrophysics Series.' The first of these books, on 'Aurora,' will be out in a few months' time and we have promises of another half-dozen subjects," adding that they "would much like to include a volume dealing with the present outlook in Cosmology, and we feel that you are the obvious person to write it."

McVittie's reply (dated 2 April 1950) contains interesting paragraphs: "Very many thanks for your letter of the 28th March and for your flattering remark that I was the person to write the Monograph in your series on cosmology. As it happens I am going to Harvard for the last three months of this year to lecture on relativity and its applications to cosmology and I have no doubt that an expanded version of these lectures would make a book of the kind you suggest. But there are two complications." One of such "complications" was not very important: his possible obligations with Methuen, for which he had written *Cosmological Theory* in 1937: In his contract he observed that it was stated that were he to write another book on the subject, then he must offer it first to Methuen for publication. The second is more interesting for my purposes:

> "The second complication is a more serious one. My feeling is that it would be a little unwise to write a book on cosmology just *before* the observations of the 200" becomes available. As you perhaps know, I view cosmology in the same way as I look upon any other branch of mathematical physics, viz. — it is a theory intended to interpret observations. The Milne+Hoyle-Bondi school look upon cosmology as an exercise in speculation and mathematical ingenuity, such observed data as there are being dealt with on the principle 'since the observations are inaccurate, it is sufficient of they do not contradict my theory too

glaringly.' I have a hunch that a lot of cosmology is going to look pretty silly as soon as the 200" comes into production and I do not want to be one of those whose faces are going to turn red."

Here, McVittie must be referring to the 200-inch Hale reflector telescope at the Palomar Observatory of the California Institute of Technology. That the telescope be named "Hale" was in honor to George Ellery Hale, who proposed its construction in 1928. Then, Hale pointed out the need for a 200-inch telescope, that will surpass the instruments at Lickm Yerkes, Hooker and Carnegie, whose "possibilities," in his opinion, "have passed out."[22] It would take, however, two decades for its construction: the dedication ceremony took place in 1948. Therefore, in 1950, when McVittie was writing to Ellison, not much definite results had come from the new powerful instrument.

Continuing with McVittie's letter, we have that he suggested that "if the Methuen difficulty can be overcome," as it finally was, he should be given the opportunity of publishing the new book in 1952 or 1953, "Meanwhile I should be getting all the theoretical side ready so that as the 200" observations appear, I could feed them in to the theoretical formula and draw the necessary conclusions. In this way, I feel sure that the work should be a worthwhile one when completed, instead of still arid exercise in speculation."

Finally McVitties's plans had to be somewhat modified because, as he wrote in a letter to Ellison (7 April 1951), during his stay in America, "apart from Stebbins & Whitford's work, nothing new seems yet to have come from the observational side and H.P. Robertson, when I saw him in Washington in the autumn, seemed doubtful whether anything would for some years." He was, "therefore more than ever doubtful about writing a book exclusively devoted to cosmology. One could, I suppose, catalogue the ad hoc theories (kinematical relativity, creation of matter, F. Jordan's numerology, etc.) and compare them with general relativity on a 'speculation' basis, using the inconclusive comparison with observation. But this seems to me to be a gloomy prospect and one which I should personally regard as a waste of time."

---

[22] Quoted in Wright, Warnow and Weiner, eds. (1972: 98), which also includes the history of the origins of this telescope and the role Hale played in it.

220 *Einstein's Relativity in Great Britain*

Instead, McVittie proposed "to write a book with some such title as 'Astronomical applications of General Relativity' deal with Einstein's theory alone. [...] The aim of the book would be to show how local gravitation (Sun, etc.) and cosmology can be dealt with by means of a single theory."

When it finally was published, in 1956 (the second edition came out in 1965; it was also translated into Russian in 1961), the book dealt in some detail with "Observational Cosmology," but only in one of the chapters, the last one (Chapter 8), mostly from the standpoint of general relativity and assuming above all that the universe can be assimilated to a perfect fluid. As it was natural of him, taking into account his previous interests, there were ample discussions about questions related to redshift, apparent magnitudes or count of galaxies and radio-galaxies. It was, in any case, a text widely used for many years, a text different from the majority of general relativity books, which did not pay the same attention as McVittie's to the relationship of theory with observation.

5.10. Settled in the United States

In spite of the advance that McVittie's new situation as professor at Queen Mary College meant, "there was no much opportunity for anything beyond routine administrative work," McVittie (ca. 1975) wrote in his autobiography; "times were difficult and the Principal of the College, Dr. (later Lord) B. Ifor Evans, and I were temperamentally unsuited to produce a fruitful collaboration." However, the College sent him as its delegate to the International Congress of Mathematicians that took place at Harvard, August 30–6 September 1950. Such an event would become decisive in his life.

Indeed, on 22 February 1950, Harold Shapley, then head of the Harvard Observatory, wrote to McVittie on the following terms:

"Dear Professor McVittie:
    We have heard that you plan to attend the Mathematical Congress in Harvard at the beginning of September of this year, and also that you might be interested in prolonging your stay in America.

We need someone to give a graduate course during the first semester of the next academic year, since Dr. Bok has gone to South Africa to work at our southern station. At a meeting of the Observatory Council yesterday, which includes the members of the Department of Astronomy in Harvard University, it was voted to ask if you could not stay on through the first semester, which begins (so far as academic duties are concerned) about the 25th of September and continues to the middle of January. We had in mind that in addition to giving one advanced course suitable to our graduate students you would also take part in the general activities of the Observatory, including the consultation with individual students about their thesis problems. By general activities of the Observatory I mean of course only the colloquia and special conferences that occur on the average perhaps once a week."

As to the nature of the course, Shapley pointed out that it could be decided later but that "possibly it might concern the general problems of cosmogony, which have not been lectured on here at the Observatory in recent years." Finally, he mentioned that Whiple, chairman of the Department of Astronomy, and Cecilia Gaposchkin "joint me in the hope that you will be able to come to the Harvard Observatory."

As it turns out, McVittie did not accept Shapley's offer and decided to stay in England. However, it did not stay long there: in 1951 he accepted an offer from the University of Illinois, in Urbana. As a matter of fact, and in an indirect way, Shapley had also something to do with the move.[23]

---

[23] In the meantime, McVittie and Shapley were in contact. Through some of the letters that have survived we can imagine that Shapley's opinion about McVittie must have been increasingly positive, as he received testimony that the British relativistic was interested and competent not only on general relativity but in astrophysics as well. Thus, on 31 May 1951 Shapley wrote McVittie: "I was glad to have your letter and note your interest in some of my galaxy work," after which he entered in technical details related to the problem of how "to get an approximately true picture of the distribution of galaxies over an area of the sky and complete to a given apparent magnitude," a problem in which he was working with Hubble ("Both Hubble and I have found...," he pointed out), although as we know he did not published anything with Hubble.

222 *Einstein's Relativity in Great Britain*

He explained his participation to McVittie in a letter dated 17 September 1951, when the negotiations were still going on:

> "I shall enclose an excerpt from one of my letters to Dean Henning Larsen of the University of Illinois. He had written to me, asking for general suggestions, primarily on the issue of closing down the department of astronomy and letting someone in mathematics do astronomy on a part-time basis. You will see from the enclosure that in a polite way I tried to point out that they were on the wrong track. In some other letters and especially in personal conversations I emphasized the point that what they should have is not a suspension of astronomical interest but a tremendous enlargement. I have tried to argue with them that there should be al least three men in a department even if there is no expensive telescopic equipment. Their big computing machine, their distinguished department of physics and chemistry and engineering, and the strong competition of the first-class universities of the Middle West, and their other assests make it seem advisable to take astronomy and astrophysics seriously.
>
> Dean Larsen tells me that my letter was reproduced and sent around the University quite a bit, for it awakened sympathetic interest and considerable understanding. It seems to me, therefore, that you going to Illinois is the first step in something that may go further. I should like to have you look into the possibility that Dr. Herget and his asteroidal computing enterprises might be transferred to Illinois. He gets very poor support from the extremely poor (financially) university in Cincinnati. But if not the highly competent Herget, possibly someone else, like Ivan King, who has a deep interest and experience in the application of computing machinery to astronomical and astrophysical problems."

Besides these possibilities, there were other "coordinating opportunities," Shapley went on, "if the Illinois program and budget can stand expansion. The borders of geochemistry, geophysics, meteorology, microwave theory and exploration — all are of astronomical interest, and should be of interest to such a great institution as the University of Illinois."

It was, indeed, a great institution, as Shapley himself had told before to McVittie, when he first heard of the possibility of his move to Illinois (11 July 1951): "Urbana is indeed a tremendous educational plant — the largest budget in America but in some respects not very lush. It probably has more members in the National Academy than any other state university (except California), and is superb in two or three fields."

In America McVittie would flourish. Not that he would become one of the uncontested leaders of the general relativity, cosmology or theoretical astrophysics there, but no doubt he was active and on the general better considered that in Britain, where, according to many indications, he seems to have had, as we have seen, a certain number of influential "enemies."[24] Although "I [am] not indissolubly tied to the U.S. — he wrote to Whittaker on 1 July 1953, commenting about the possibilities of being the successor of Max Born in Edinburgh — I have found here far greater opportunities for research than in London." In 1952, he was elected to the American Astronomical Society, serving as its Secretary during the period 1961–1969, a position that entailed travel to many cities of the continental United States and Canada, with long-range excursions to Fairbanks, Alaska and Hawaii.[25] He had also a long association with Commission 28 (Galaxies) of the International Astronomical Union, in which he was successively Secretary (1958–1964), Vice-President (1964–1967) and President (1967–1970).

5.11. McVittie and Radio Astronomy at Urbana University

Among the opportunities McVittie found in America, there was an easier access to observations. We have seen repeatedly that his scientific and philosophical outlook was such that he was not satisfied with a purely theoretical approach to gravitation and cosmology. However, in Britain,

---

[24] These comments do not mean that he had not receive any recognition while in Britain. Thus, in 1931 he was elected a Fellow of the Royal Astronomical Society, of whose Council he served during 1942–1946. He was elected also Fellow of the Royal Society of Edinburgh (1943), and of the Royal Meteorological Society (1948), as recognition of the meteorological work he had done during War World II (see on this regard Knighting 1990 and Hide 1990).

[25] Some of McVittie's activities as Secretary of the American Astronomical Society are mentioned in DeVorkin, ed. (1999).

224   *Einstein's Relativity in Great Britain*

observational possibilities remained distant. No so in Illinois. Here is how he referred in his autobiography (McVittie ca. 1975) to what happened there:

> "At the instigation of Professor Edward C. Jordan, shortly to become the head of the Electrical Engineering department at Illinois, I was sent by the University in 1954 to conferences in Washington on radio astronomy. Eventually the plan materialized in the National Radio Astronomy, at Green Bank, W. Virginia. However, at the Washington conference there was a group to which I belonged that emphasized the necessity of smaller radio astronomy projects at Universities as well. In these, young radio astronomers would be trained. In pursuing this idea, Jordan and I appointed in 1956, George W. Swenson, Jr., an electronics expert specializing in antenna design, to half-time professorships in each of our departments. At that time extragalactic radio sources would provide a good criterion for selecting the appropriate model of the universe. I urged Swenson to plan an instrument suitable for survey work and able to detect faint, and therefore presumably very distant, radio sources."

The project was founded by the Office of Naval Research. Construction of a 600×400 ft. parabolic cylinder dish, fixed to the ground, which operated at 610.5 MHz, began in September 1959. Late in 192, the instrument came in operation, at a location which they named Vermilion River Observatory, 30 miles east of Urbana campus. However, different problems, including the cutting back of support by the Office of Naval Research and the Federal Government, and Swenson's lack of interest in the reduction of the survey data from the instrument, led to the fact that the sky survey planned remained uncompleted when the instrument was retired by June 1970. But this is, however, another history. What I want to emphasize is how different was for George McVittie the American setting from the British. The frontiers between the theoreticians and the experimentalists were not as strict as, in several aspects, were in Britain, and in general in Europe, a fact that has been pointed out by Michael Eckert (1996) with respect to solid state physics in Germany and the United States. Eckert argued that suck lack of — or difficulty in — communications and interchanges explain the ultimate failure of the Sommerfeld school on becoming the world leader in the new field, then emerging, of solid state physics. I guess that something similar can be said about important areas of gravitation physics, astrophysics and cosmology.

A proof, albeit indirect, of such differences is what McVittie wrote to Allan Sandage on 21 December 1966:

"Dear Allan,

I hope to see you at the AAS meeting next week but meanwhile I wanted to raise a point with regard to a preprint of an article entitled 'Radio Astronomy and Cosmology' by P.A.G. Scheuer for Vol. IX of 'Stars and Stellar Systems" of which you are the editor. I wonder if you, as editor, think it quite fair that the work of my pupil and myself — and the work done at the VRO [Vermilion River Observatory] — should be totally ignored? The papers I have more particularly in mind are

G. C. McVittie, Austr. J. of Physics, $\underline{10}$, 331, 1957.

G. c. McVittie and R. C. Roeder, Ap. J. $\underline{138}$, 899, 1963.

G. C. McVittie and L. Schusterman, A. J., $\underline{137}$, 1966.

Yet I observe that Scheuer gives numerous references to the work of W. Davidson who always seems to me to re-write my papers in an intricate notation and publish them in M. N. [*Monthly Notices*] Of course, I know that I am not persona grata in Britain and particularly in Cambridge where our work at VRO on surveys is regarded as an unwarranted intrusion on their private preserve! But I wondered if you would feel quite happy about this attitude appearing in an American publication. I might add that the second of the above papers drew favorable comments from Minkowsi at Padua in 1964."

5.12. McVittie and the Royaumont General Relativity and Gravitation Congress (GR 2)

The documents conserved among the McVittie's papers include several reports in which the British relativist informed of his travels and meetings to which he attended. To finish the present paper I will consider one of those reports, the one in which McVittie reviewed the Colloquium dedicated to Relativistic Theories of Gravitation held at Royaumont, near Paris, from June 18 till August 8, 1959.

That meeting is especially worth our attention because it was one of the first General Relativity and Gravitation Conferences, which came to be known as GR. Following the nomenclature introduced by André Mercier (1992), the Royaumont Conference would be GR2 (Royaumont 1962), after GR0, the "Jubilee of Relativity Theory" meeting held in Bern, July 11–16, 1955 (Mercier and Kervaire, eds. 1956), and GR1, the Conference on the Role of Gravitation in Physics, held at the University of North Caroline, Chapel Hill, January 18–23, 1957 (DeWitt, ed. 1957). It was, therefore, one of meetings which contributed most to the institutionalization of general relativity. And it was also the first of the GR Congresses to which McVittie attended.

As to the report mentioned, we have that McVittie dated it on August 26, 1959, and entitled it: "Report on Travel sponsored by Office of Naval Research (Electronics Branch, Contract ONR 1834 (22)), June 18–August 8, 1959". Here it is what it said:

> "(1) Colloquium on 'Les Theories Relativistes de la Gravitation', Royaumont, near Paris, France, June 21–27, 1959.
>
> This week-long symposium was attended by some 100 persons drawn from 14 countries. The U.S contingent was the largest, 40 in number, followed by France with 21. Three came from the USSR, four from Poland and three from E. Germany.
>
> Much time was devoted to the concept of energy in general relativity. Since no tensor definition of energy is to be found in this theory, attention was concentrated on non-tensorial definitions that seemed plausible. Of these the most interesting was that of P. A. M. Dirac (Cambridge, England) who suggested that energy was an integral of the equations of motion which had the correct physical dimensions and was useful in studying the equations.
>
> Gravitational waves were much discussed. Exact solutions of Einstein's equations certainly exist in which gravitational effects can be regarded as propagated with the speed of light. Little interest was evidenced in the astronomical consequences of such a finite speed of propagation for gravitation or in its physical detection by experiments. There was however one paper by J. Weber (Univ. of Maryland, USA) on the possibility of detecting gravitational waves experimentally. One proposed method was to observe the relative motion of masses which interact with a gravitational wave. The second was to employ the strains set up in a solid by such a wave. It was not clear that the effects to be

observed would be distinguishable from the 'noise' inherent in the proposed experimental methods.

The quantization of general relativity received some attention though no significant progress appears to have been made since the first of these conferences in 1950. With the death of Einstein, the search for a unified field theory of gravitation and electromagnetism has apparently faded into the background.

The group of French mathematicians whose leader is A. Lichnerowicz (Institut H. Poincaré, Paris) produced much interesting pure mathematics on the propagation of discontinuities in the space-times of general relativity and on the properties of various tensors that can be defined therein.

Cosmology was treated in two papers only: one of these was the lecture by the present writer (copy attached) which dealt mainly with the distribution in space of faint radio sources. The other was by D. N. Sciama (London, England) on the present observational situation in cosmology. These two papers, together with another one having astronomical implications by A. H. Taub (U. of Illinois, USA) on small motions of spherically symmetric distributions of matter, were discussed at an extra session. The writer acted as chairman of this session."[26]

The conference was well attended indeed: 119 participants, McVittie among them, and with the noted mathematician and contributor to general relativity André Lichnerowicz presiding.

As McVittie stated, there were a significant number of contributions devoted to the concept of energy on relativity: Those, for instance, of Christiaan Moller, Felix Pirani, Paul Dirac and Stanley Deser. Thus Moller's contribution (which opened the volume of proceedings) was

---

[26] Next, McVittie summarized visits he had made to the Max-Planck Institut für Physik und Astrophysik, Munich, June 29–July 2, where he delivered a lecture, and to the Jodrell Bank Experimental Station, near Manchester, July 13–16. About the last he wrote: "I was shown over the facilities at this Observatory, particularly the 250-foot dish, and found the whole installation even more impressive than I anticipated. Some 200 to 250 extragalactic radiosources have been measured for angular diameter with the 250-ft. dish. The highlight of the visit was a long discussion with R. C. Jennison on his investigations of the radio-source Cygnus A. He has shown that the source consists of two sources of almost equal intensity, with the pair of colliding galaxies between them..."

228 *Einstein's Relativity in Great Britain*

entitled "The energy-momentum complex in general relativity and related problems". Moller's opening words serve well to illustrate the nature of the problem (Moller 1962: 15): "Within the framework of Einstein's theory of gravitation it is possible to define a large number of algebraic functions of the field variables which satisfy 'conservations laws,' and the problem arises how to determine which of these functions represent quantities with a physical meaning."

Gravitational radiation was, as McVittie also mentioned, another of the topics dealt with in several interventions: like André Lichnerowicz, his student the Spanish physicist who settled in Paris, Luis Bel, of "Bel-Robinson tensor" fame, Vladimir Fock, and, on the experimental side Joseph Weber, whose talk was "On the possibility of detection and generation of gravitational waves."

We know that during decades, general relativity was dominated by mathematics. It was still so, generally, when GR2 was held, and this is something that can be appreciate by simply its proceedings. Precisely because that, surely McVittie ought not to have been very satisfied with the Royaumont congress: too many mathematical papers were read there. Papers like: Jürgen Ehlers, "Transformations of static exterior solutions of Einstein's gravitational field equations into different solutions by means of conformal mappings;" Yves Thiry, "Sur les théories pentadimensionnelles;" Jean-Marie Souriau, "Relativité multidimensionnelle non stationnaire;" Olivier Costa de Beauregard, "Quelques remarques d'analyse dimensionnelle pouvant intéresser les futures théories unitaires;" Cécile DeWitt, "Grandeurs relatives a plusieurs points. Tenseurs generalizes;" David Finkelstein and Charles Misner, "Futher results in topological relativity;" or Roger Penrose, "General relativity in spinor form."

Apparently, he should have been happier with Peter Bergmann and Arthur Komar's contribution: "Observables and commutation relations observables et regles de commutation." The appearance of the word "observable" must have sounded well to McVittie's ears, even though the paper was in fact dedicated to a highly theoretical subject: the quantization of general relativity. What, however, meant Bergmann and Komar for "observables"? In their words (Bergmann and Komar 1962: 313): "In principle, invariant quantities represent intrinsic properties of a physical situation, properties that are independent of the (equivalent) modes of description. If our theory deals of physically meaningful quantities, invariants should possess expectation values; and to the extent that in our theory observable quantities can be predicted, invariants should be predictable from sufficiently complete data given on one space-like hypersurface. This, then, is the motivation for talking of observables and for proposing to formulate the whole physical theory as far as possible in terms of observable theory." And then they added: "In general relativity many observables are constants of the motion outright. From any that are not we can construct constants of the motion, by expressing the value of an observable at the coordinate time x_0^0 in terms of observables at another (variable) coordinate time x^0." Later on, they dedicated themselves to the hard task of constructing observables.

However, it seems that, not surprisingly, McVitte was not completely satisfied with their approach, and one can perceive a point of irony in his intervention (Royaumont 1962: 323) at the discussion that followed the presentation of the paper (made by Bergmann): "I find absolutely convenient to consider the observables as essential in general relativity. An example is provided by the determination of the distance by means of the apparent value of a luminous extended source, in which case an operational prescription of the measure can be done. Could Mr. Bergmann give us an example of observable: (a) which be the natural result of the association of general relativity and the quantum theories; (b) and for which it could be possible to indicate an operational measure process?". According to the proceedings, his question remained unanswered.

As to his own contribution to the congress (entitled "Cosmology and the interpretation of astronomical data" [McVittie 1962b]), in it he discussed on one side three cosmological problems: the applicability of uniform cosmological models to the observed universe, the status of the problem of the expansion of the universe, and the distribution in space of extragalactic Class III radio sources. At the same time, he considered the applicability to those problems of different gravitational theories: general relativity, steady state, and kinematical relativity. "The formulae through which a comparison of theory and observation is made," he stated (McVittie 1962b: 253), "are indeed for the most part so nearly identical in form that all three theories can be treated together."

230　*Einstein's Relativity in Great Britain*

In his conclusions, McVittie (1962b: 265) considered several questions, among them the use of the cosmological constant, that he still defended ("attention should be concentrated on those [uniform cosmological models] having hyperbolic space and a negative cosmic constant"), but it was especially against the steady-state cosmology that he addressed his critics:

> "I think it is illusory to claim, as is done in the steady-state theory and in kinematical relativity, that one single highly specialized model universe can be chosen to represent the observed universe...
>
> Supporters of the steady-state theory suggest that new kinds of observations are needed [F. Hoyle, 'Paris Symp. on Radio Astronomy', p. 529. University Press, Stanford (1959)] in order to 'test' their theory as against general relativity. The observers [M. Ryle, *Proc. Roy. Soc. A, 248*, 289 (1958); J.L. Pawsey, Trans. I.A.U. (Moscow 1958), *10* (in press). Report of Commission 40] also appear to believe that the question is still open. In fact, it has been known since 1956 that the steady-state theory predicts the wrong sign for the acceleration parameter, We have also pointed out that the predicted average density of matter is rather too high. And lastly, one result of the present paper has been to show that the steady-state theory fails to reproduce the empirical law of distribution of Class II radio sources." And then he concluded: "In view of these considerations, it is not clear how further 'tests' could validate the steady-state theory. Its model universe simply does not agree with observation whereas, as we have seen, certain general relativity model universes do."

Besides reading a paper, McVittie participated in a number of discussions, always faithful to his critical personality that brought him a not small number of enemies all through his life. Thus, during the discussion after Joshua Goldberg's intervention ("Conservation laws and equations of motion"; Goldberg 1962), which was of course mainly mathematical, McVittie made a question which was perfectly fitted to his own philosophy of science (Royaumont 1962: 43): "Would you like to tell us which is the reason of this work? Is it conceived as an exercise of analysis or perhaps it can through some light on specific physical problems? For instance, does it make easier the solution to the two bodies? If this is too difficult, could one in this manner have more information about the motion of Mercury than the one that can be obtained with a direct solution?"

Goldberg's reply deserves to be quoted, at least in part (Royaumont 1962: 43): "The aim of one theory of motion in general relativity is not to obtain more efficiently, or with more precision, the perihelion precession, but to explore till what point the general relativity theory contains those predictions. There are a certain number of fundamental questions in the [General Relativity] theory of motion which are not yet completely clarified. Surely, to attach an observation or a given experience to the theory, or vice versa, we must ask ourselves, one time after another, till point the theory predicts the motion of material bodies in a gravitational field."[27]

We know what he really thought of the Royaumont congress through a letter he wrote to his British colleague Gerald J. Whitrow, on August 18, 1959: "The Paris relativity conference was indeed boring in parts: endless re-hashing of 'energy' considerations, much talk about gravitational waves, hankering after unification of quantum mechanics and general relativity, etc. But the French group under Lichnerowicz produced some very elegant pure mathematics." And in a new note against the steady-state theory: "If such a secular effect [a greater rate for the formation of rich clusters than the one predicted by the steady-state theory] exists, it would provided yet another observational argument against the steady-state theory. Not that such evidence will have any effect on the 'cosmological' climate in England! The steady-state boys, Bondi, Hoyle, Pirani, etc., are far too good publicists to let such points (or those listed at the end of my Paris paper) throw them off their stride."

Such was that man who, fittingly, was called "the uncompromising empiricist."

---

[27] There were other participants whose interventions were dedicated to conservation laws in general relativity; for instance, John L. Synge and Andrzej Trautman.

Stephen Hawking, January 1999. NASA StarChild Learning Center (Public Domain).

Chapter 6

The Renaissance of Relativity in Great Britain: From Coordinates to Global Space-Time and Black Holes

History, of science or of whatever other discipline, is like a continuous thread dotted with knots which sometimes grow introducing substantial novelties. The history of Einstein's relativity in Great Britain, and subsequently everywhere, is not foreign to such sort of developments. The most notorious of those "knots", the analysis of the space-time structure from a global viewpoint, was the one introduced by Roger Penrose, and following his steps by Stephen Hawking. However, it would be a gross mistake to reduce the history of relativity in Britain to that novel approach forgetting the contributions of the group assembled around Hermann Bondi at King's College London, where, as we saw in Chapter 5 also spent some time McVittie. The great difference between the approach of Penrose and others, and that of the Bondi's group was that in the latter case the coordinate-style dominated their researches.

6.1. General Relativity at King's College London

The first thing to say is that general relativity was "housed" at the Department of Mathematics of King's, a feature not uncommon as far as how general relativity was cultivated during many years, in Great Britain

235

236 *Einstein's Relativity in Great Britain*

as well as elsewhere, until the arrival of new and powerful astrophysical and cosmological techniques.

As David Robinson (2019), who joined the Department of Mathematics at King's in 1970, remaining there until 1996 being appointed Emeritus professor in 2001, has explained, in the decade after the end of the Second War World there was a slow but steady renewal of interest in general relativity and King's was one of the centers where such interest were manifested. It was in 1954 when Bondi arrived to replace George Temple as the professor of Applied Mathematics and he quickly formed a research group consisting initially of himself, Clive Kilmister (1924–2010) who was already at King's, and Felix Arnold Edward Pirani (1928–2015), a Canadian — he had studied at the University of Western Ontario and got his doctor degree at the Carnegie Institute of Technology in 1951 with a thesis entitled *On the quantization of the gravitational field of general relativity*, with Alfred Schild as supervisor — who arrived the following year.

The situation of the field of general relativity at that time was reasonably summarized by not less than the great Richard Feynman when he attended the International Conference on Relativistic Theories of Gravitation held from 25–31 July, 1962 at Warsaw and Jablonna (Poland).[1] In an often-quoted letter to his wife, he wrote (Feynman, Michelle, ed. 2005: 137–138):

> "I am not getting anything out of the meeting — I am earning nothing. This field (because there are no experiments) is not an active one, so few of the best men are doing work in it. The result is that there are host (126) of dopes here — and it is not good for my blood pressure — such inane things are said and seriously discussed — and I get into arguments outside the formal sessions — say at lunch — whenever anyone asks me a question or starts to tell me about his 'work.' It is always either — (1) completely un-understandable, or (2) vague and indefinite, or (3) something correct that is obvious and self-evident worked out by a long and difficult analysis and presented as an important discovery, or (4) a claim,

---

[1] This conference became known as GR3 ("General Relativity 3"), after Bern (GR0, 1955), Chapel Hill (GR1, 1957), and Royaumont (GR2, 1959). GR4 was held at Imperial College, London, in 1965, and it was Bondi who edited the proceedings,

based on the stupidity of the author that some obvious and correct thing accepted and checked for years is, in fact, false (these are the worst — no argument will convince the idiot), (5) an attempt to do something probably impossible, but certainly of no utility, which, it is finally revealed, at the end, fails (dessert arrives, is eaten), or (6) just plain wrong. There is a great deal of 'activity in the field' these days, but this 'activity' is mainly in showing that the previous 'activity' of somebody else resulted in an error or in nothing useful or in nothing promising. It is like a lot of worms trying to get out of a bottle by crawling all over each other. It is not that the subject is hard; it is just that the good men are occupied elsewhere. Remind me not to come to anymore gravity conference."

To understand fully what Feynman said, we must take into account that he formed part of a scientific culture, that of searching a quantum electrodynamics, in which experimental data were essential. In his now classic study of the genesis of quantum electrodynamics, Silvan Schweber (1994: xii) explained this characteristic as follows:

"Many of the advances during the 1945–1950 period, especially on the experimental side, were made in the United States. One of the principal reasons for this was that the United States did not suffer the devastation that World War II had inflicted on Great Britain, continental Europe, and Japan. But there were other factors at play, reflecting the institutional settings in which physics was practiced in the United States and the great technical advances made during World War II. In the United States, theory and experiments were always housed under the same roof in one department. And perhaps more so than everywhere else, in the United States physics was about numbers, and theories were deemed to be algorithms for getting the numbers out. In the wartime laboratories an entire generation of young theorists was raised with this attitude. World War II played an important role in the maturing of the American physics community. It is thus not surprising that the physicists' wartime efforts and the influence of these experiences on the subsequent developments should receive attention there."

General relativity could not follow those procedures simply because there were very few, if any, experimental data.

Independently of Feynman's assessment, that conference was important because it was the first time that physicists from the West and the East were able to meet personally. Among the 114 participants, there were distinguished scientists like Dirac, Wheeler, Sciama, Bergmann, Chandrasekhar, DeWitt, Lichnerowicz, Møller, Rosenfeld, Ginzburg, Fock, Petrov, Ivanenko, Mandelstam, Plebanski, Misner, Weber, Synge, Bondi, who spoke about "Radiation from an isolated system" and "The steady state universe", Penrose ("The light cone at infinity"), and Feynman himself ("The quantum theory of gravitation"). And Bondi, and well as Pirani profited from it, as well as from the previous GR conferences, those in Bern and Chapel Hill.

Feynman delivering his lecture at Jablonna (photograph by Yurij Vladimiov).

Paul Dirac with Richard Feynman (photograph by Marek Holzman).

Hermann Bondi with Peter Bergmann (photograph by Marek Holzman).

As I.W. Roxburgh (2007: 54) pointed out in the obituary of Bondi he wrote for the Royal Society, "in the mid 1950s, there was debate as to whether Einstein's general theory of relativity predicted the existence of gravitational waves", and at a meeting in 1955 in Bern, Marcus Fierz took Bondi "on one side and said, 'the problem of gravitational waves is ready for solution and you are the person to solve it'. Thus gravitational waves became the focus of research at King's. In January 1957 Bondi and Pirani participated in [already mentioned in Chapter 5] conference on General Relativity at Chapel Hill, NC, USA; this received some support from the US Air Force, which had started a program funding research in General Relativity. Discussions with Joshua Goldberg led to a contract with King's College to support Bondi's group. During the 1950s and 1960s King's became a world center for research in relativity, establishing close links with Leopold Infeld's group in Poland and attracting many students and visitors including Leslie Marder, Peter Szekeres, Andrzej Trautman, Alfred Schild, David Robinson, Roger (later Sir Roger) Penrose (FRS 1972), Goldberg and many others. The group's research covered gravitational radiation, epistemological foundations, high gravitational potentials, slowly changing fields and special solutions. Their work on gravitational

240 *Einstein's Relativity in Great Britain*

radiation led to a series of papers "Gravitational waves in General Relativity" published in Proceedings of the Royal Society A from 1958 to 2004; some were by Hermann himself, some were by his students, and others were written with collaborators or by visitors to King's. Two of these papers deserve special mention. In their 1959 paper III, 'Exact plane waves' by Bondi, Pirani and Robinson, they gave the first exact solution for such waves, thus finally demonstrating that General Relativity did indeed predict gravitational waves. In their 1962 paper VII, 'Waves from axi-symmetric isolated systems', by Bondi, van der Burg and Mezner, they obtained a clear understanding of the transport and reception of energy in such waves, introducing the 'Bondi radiation coordinates', the 'Bondi mass', the 'news function' and the Bondi–Metzner–Sachs group, thereby laying the foundation for much future work on the subject. Hermann considered this his most important paper."[2]

The subject of gravitational waves in general relativity had interest and also a long tradition, dated back to Einstein's papers in 1917 and 1919, Einstein and Rosen in 1937, and Rosen alone in 1937, but on the whole it shared what Feynman criticized: too mathematical and foreign to experimental verification, a characteristic which also appear in most Bondi's papers, for example in "Spherically symmetric models in general relativity", "Gravitational bounce in general relativity" (Bondi 1947, 1969), or "Energy transfer by gravitation in Newtonian theory" (Bondi and McCrea 1960). One may wonder if Bondi himself would have not realized such too mathematically oriented researches, which little added to the knowledge of reality, and decided to switch to the administration, in which he carried on a very successful career being (1967–1970) Director General of ESRO (European Space Research Organization), Chief Scientist of the British Ministry of Defense (1971–1977), Chief Scientist of the Ministry of Energy (1977–1980), and Chairman and Chief Executive of the National Environmental Research Council (1980–1984), a position which he combined since 1982 with the Master of Churchill College, Cambridge, which he retained up to 1990.

---

[2] See Bondi, Pirani and Robinson (1959) and Bondi, Van der Burg and Metzner (1962). His last paper in the series, XVI, *Standing waves*, was published in 2004 (Bondi 2004).

Felix Pirani also shared such approach to general relativity. Today, his most appreciated works was the article he published in 1956 in *Acta Physica Polonica*, "On the physical significance of the Riemann tensor" (Pirani 1956).

6.2. Penrose, Hawking and the global structure of space-time

The approaches and treatment of the general relativity in Britain, as elsewhere, had not experimented with real changes since Einstein introduced it, and Eddington exposed in his masterful *The Mathematical Theory of Relativity* (1923). With the only exception of Alfred Robb, it was an approach based on the use of Riemannian geometry and coordinates. This changed thanks to a mathematician who became attracted by Einstein's theory: Roger Penrose (b. 1931) and the subsequent works of Stephen Hawking (1942–2018). To understand how a very promising young mathematician changed field, let Penrose (2011: xii) speak:

"As an undergraduate, I studied Mathematics at University College London, from 1951 to 1954; then I went to St. John's College Cambridge to do research in algebraic geometry, initially under William Hodge and subsequently John Todd. In my same year, under Hodge, was Michael Atiyah, who was an inspirational, though somewhat intimidating colleague. [...]

Yet the colleague who had the greatest influence on the development of my research during this period, and for many years later, was the cosmologist Dennis Sciama. Even before studying at Cambridge, I had a chance meeting with him during a visit to my brother Oliver in Cambridge in which I raised a question that had worried me in relation to the inspiring series of BBC talks given by Fred Hoyle in 1951 (The nature of the universe. A series of broadcast lectures). On the strength of this encounter (at which I used a space-time diagram with tilted cones to make my point), Dennis had evidently formed the opinion that it would be worth while to interest me more in physics, even trying, later, to persuade me to switch my topic of research to cosmology. I was then too committed to my many pure mathematics activities, but I learnt a great deal of physics from Dennis, and of the excitements involved in striving

242 *Einstein's Relativity in Great Britain*

to uncover secrets of the universe. Another influential colleague was Felix Pirani, from whom I learnt much about developments in general relativity.

Pivotal were the superb lecture courses of Bondi, on general relativity, and Dirac, on quantum mechanics. On the pure-mathematical side, I recall particularly an elegant information-packed algebra course by Philip Hall."

Penrose gave great credit to Dennis Sciama (1926–1999), who was also, as I will soon explain, very influential in directing the career of Hawking. In their obituary for the Royal Society, George Ellis and Roger Penrose (2010: 401) made a short but just appreciation of Sciama's contributions: "The quarter-century period starting in the mid-1950s is sometimes referred to as 'the renaissance of general relativity and cosmology'. Many contributed to this activity, but in the UK a major factor was the guidance and leadership provided by the warmly enthusiastic personality of Dennis Sciama, whose broad knowledge and deep insights greatly inspired many in their individual achievements." A simple, but revealing, measure of his influence is that he was the PhD supervisor of, among others, George Ellis (1964), Stephen Hawking (1966), Brandon Carter (1967), Martin Rees (1971), Gary Gibbons (1973), John Barrow (1977), Philip Candelas (1977), David Deutsch (1978).

Sciama had obtained his doctoral degree with no less that Paul Dirac as first advisor and Hermann Bondi as second, with a thesis dealing with one of his favorite themes, *Mach Principle and Inertia* (Sciama 1953), a subject to which he returned several times; thus in a book published in 1959, and dedicated to "Hermann Bondi, Tommy Gold and Fred Hoyle", he wrote (Sciama 1959: 84–85): "According to Mach, a body has inertia because it interacts in some way with all the matter in the universe. [...] My own view is that Mach is right."[3] "Dennis Sciama — we read in an article about 'cosmology's rebirth' (De Swart 2020: 266) — who is often recognized as one of the fathers of modern cosmology, was central in the revival of work on Mach's principle. His doctoral work under the supervision of Paul Dirac resulted in a novel take on the 'origin of inertia,' in which he proposed a type of long-range interaction with distant matter.

---

[3] He gave a detailed discussion of Mach Principle in his book *The Physical Principles of General Relativity* (Sciama 1969).

The Renaissance of Relativity in Great Britain   243

Mach's principle was put high on the list of newly developing research agenda of gravitational physics."

Though Sciama's influence was especially felt in England, mainly in Cambridge (1950s and 1960s) and afterwards in Oxford (1970s and early 1980s), he travelled and extensively. He taught at Cornell University, Harvard University, the University of Texas at Austin, the Trieste International School for Advanced Studies, and the Scuola Normale Superiore of Pisa. But what interest here now is the influence he exerted in Stephen Hawking, who referred to this in his autobiography, *My Brief History* (Hawking 2013: 41–42):

"I arrived in Cambridge as a graduate student in October 1962. I had applied to work with Fred Hoyle, the most famous British astronomer of the time, and the principal defender of the steady-state theory. I say astronomer because cosmology was at that time hardly recognized as a legitimate field. That was where I wanted to do my research, inspired by having been on a summer course with Hoyle's student Jayant Narlikar. However, Hoyle had enough students already, so to my great disappointment I was assigned to Dennis Sciama, of whom I had not heard.

It was probably for the best. Hoyle was away a lot, and I wouldn't have had much of his attention. Sciama, on the other hand, was usually around and available to talk. I didn't agree with many of his ideas, particularly on Mach's principle, the idea that objects owe their inertia to the influence of all the other matter in the universe, but that stimulated me to develop my own picture.

When I began research, the two areas that seemed most exciting were cosmology and elementary particle physics. The latter was an active, rapidly changing field that attracted most of the best minds, while cosmology and general relativity were stuck where they had been in the 1930s. [...]

But I felt that the study of elementary particles at that time was too like botany. Quantum electrodynamics — the theory of light and electrons that governs chemistry and the structure of atoms — had been worked out completely in the 1940s and 1950s. Attention had now shifted to the weak and strong nuclear forces between particles in the nucleus of an atom, but similar field theories didn't seem to work to explain them. Indeed, the Cambridge school, in particular, held that there was no underlying field theory. Instead, everything would be determined by unitarity — that is, probability conservation — and certain

244 *Einstein's Relativity in Great Britain*

characteristic patterns in the scattering of particles. [...] Cosmology and gravitation, on the other hand, were neglected fields that were ripe for development at that time. Unlike with elementary particles, there was a well-defined theory — the general theory of relativity — but this was thought to be impossibly difficult. People were so pleased to find any solution of the Einstein field equations that describe the theory that they didn't ask what physical significance, if any, the solution had. This was the old school of general relativity that Feynman had encountered in Warsaw. Ironically, the Warsaw conference also marked the beginning of the renaissance of general relativity, though Feynman could be forgiven for not recognizing it at the time. A new generation entered the field, and new centers of the study of general relativity appeared. Two of these were of particular importance to me. One was located in Hamburg, Germany, under Pascual Jordan. I never visited it, but I admired the elegant papers produced there, which were such a contrast to the previous messy work on general relativity. The other center was at King's College, London, under Hermann Bondi. Because I hadn't done much mathematics at St. Albans or in the very easy physics course at Oxford, Sciama suggested I work on astrophysics. But having been cheated out of working with Hoyle, I wasn't going to study something boring and earthbound such as Faraday rotation. I had come to Cambridge to do cosmology, and cosmology I was determined to do. So I read old textbooks on general relativity and traveled up to lectures at King's College, London, each week, with three other students of Sciama's. I followed the words and equations, but I didn't really get a feel for the subject."

Hawking submitted his PhD dissertation, entitled *Properties of Expanding Universes* on February 1966. Its abstract reads as follows:

"Some implications and consequences of the expansion of the universe are examined. In chapter 1 it is shown that this expansion creates grave difficulties for the Hoyle-Narlikar theory of gravitation. Chapter 2 deals with the perturbations of an expanding homogeneous and isotropic universe. The conclusion is reached that galaxies cannot be formed as a result of the growth of perturbations that were initially small. The propagation and absorption of gravitational radiation is also investigated in this approximation. In chapter 3 gravitational radiation in an expanding universe is examined by a method of asymptotic expansions. The 'peeling-off' behavior and the asymptotic group are derived. Chapter 4 deals with the occurrence of singularities in cosmological models. It is shown that a singularity is inevitable provided that certain very general conditions are satisfied."

It was that last chapter that marked Hawking's future career. Especially important in this sense was Section 4, "Singularities in Inhomogeneous models", of that chapter. There, Hawking (1966) wrote:

"Lifshitz and Khalatnikov ["Investigations in relativistic cosmology," *Advances in Physics 12*, 185–249 (1963)] claim to have proved that a general solution of the field equations will not have a singularity. Their method is to contract a solution with a singularity which they claim is representative of the general solution with a singularity, and then show that it has one fewer arbitrary function than a fully general solution. Clearly their whole proof rests on whether their solution is fully representative and of that they give no proof. Indeed it would seem that it is not representative since it involves collapse in two directions to a 1-surface whereas in general one would expect collapse in one direction to a 2-surface. In fact their claim has been proved false by Penrose ["Gravitational collapse and space-time singularities," *Physical Review Letters 14*, 57–59 (1965)] for the case of a collapsing star using the notion of a 'closed trapped surface'. A similar method will be used to prove the occurrence of singularities in 'open' universe models."

What it is important is to point out that it was Roger Penrose (1965) which opened the field, showing that Lifshitz and Kalatnikov were wrong: small departures from spherical symmetry will not prevent a singularity. Instead of centering his analysis in Einstein's field equations of general relativity, Penrose based his work on global methods, the study of space-time causal structure, and energy inequalities.

George F. R. Ellis, a native of South Africa who after studying in the University of Cape Town, where he graduated in 1960, went to St. Johns College, Cambridge, where, as already said, got his Ph.D. degree with Sciama, explained how was Penrose's 1965 article received (Ellis 2014: 405):

"After the publication of this paper [Penrose's] in January 1965, the members of Dennis Sciama's general relativity group in the Department

246   *Einstein's Relativity in Great Britain*

of Applied Mathematics and Theoretical Physics at Cambridge University (particularly Stephen Hawking, myself, and Brandon Carter) hurriedly tried to learn the new methods that Penrose had introduced. We were assisted in this by discussions with Felix Pirani and the group at King's College, London; with John Wheeler and Charles Misner, who visited Cambridge from the USA for an extended period; and with Roger Penrose and Bob Geroch, who was visiting Penrose at Birkbeck College, London.[4] In particular we had a one day seminar in Cambridge attended by the members of the King's College group, where I and Brandon Carter summarized our understandings of the ingredients of Penrose's theorem. Part of the requisite material was the coordinate-free approach to differential geometry that I had introduced to the Cambridge group in lectures in late 1964 after learning it from the excellent book by Helgason [*Differential geometry and symmetric spaces*, 1962] and Cohn [*Lie Groups*, 1957]; part was the material on focusing of timelike and null geodesics that we had learnt from Schücking, Ehlers, and Sachs [Ehlers, 1961; Jordan, Ehlers, and Sachs, 1961] and from Newman and Penrose [1962]; part was the study of causal structure we had heard in seminars given by Penrose [1963]; and part was the new material introduced by Penrose [1965] on how the focusing of null geodesics at caustics and folds bounded causal domains."

In his autobiography, Hawking (2013: 63) acknowledged the influence that Penrose's approach had in him:

"But it wasn't clear that the solution they [Lifshitz and Kalatnikov] found was the most general one. Roger Penrose introduced a new approach, which didn't require solving Einstein's field equations explicitly, just certain general properties, such as that energy is positive and gravity is attractive. Penrose gave a seminar on the subject at King's College, London, in January 1965. I wasn't at the seminar, but I heard about it from Brandon Carter, with whom I shared an office in

---

[4] Geroch's lectures were used by Martin Walker to write a report, *Spacetime Structure from a Global Viewpoint* (43 pages), which was edited by the Department of Mathematics, King's College London. Walker, an active relativistic, would make his academic home at the Max Planck Institute für Physik und Astrophysik in Munich.

Cambridge's new Department of Applied Mathematics and Theoretical Physics (DAMTP) premises in Silver Street.[5]

At first I couldn't understand what the point was. Penrose had shown that once a dying star contracted to a certain radius, there would inevitably be a singularity, a point where space and time came to an end. Surely, I thought, we already knew that nothing could prevent a massive cold star from collapsing under its own gravity until it reached a singularity of infinite density. But in fact the equations had been solved only for the collapse of a perfectly spherical star, and of course a real star won't be exactly spherical. If Lifshitz and Khalatnikov were right, the departures from spherical symmetry would grow as the star collapsed, and would cause different parts of the star to miss each other, thus avoiding a singularity of infinite density. But Penrose showed they were wrong: small departures from spherical symmetry will not prevent a singularity."

And in the next five years, Penrose, Hawking and the American theoretical physicist Robert Geroch developed the theory of causal structure in general relativity. "It was a glorious feeling having a whole field virtually to ourselves," wrote Hawking (2003: 111) in the book celebration of his 60th birthday.

Proof that the new approach did not pass unnoticed is that when in 1966 it was announced the prestigious Adams Prize, awarded each year by the Faculty of Mathematics at the University of Cambridge and St John's College, the topic set was "Geometric Problems of Relativity, with special reference to the foundations of general relativity and cosmology."[6] And both, Penrose and Hawking, submitted essays. The adjudicators of that

---

[5] Brandon Carter, an Australian native, was an active member of the Cambridge relativity group. One of his most important works was made together with Werer Israle and Stephen Hawking: a partial proof of the no-hair theorem in general relativity. He settled finally in France, as member at the Laboratoire de l'Univers et de ses Théories of the CNRS, at Meudeon.

[6] The prize had been established university in 1848 to commemorate Adams' discovery of the planet Neptune. Among the winners were James Clerk Maxwell (1857), J.J. Thomson (1882), John H. Poynting (1893), Joseph Larmor (1899), Geoffrey I. Taylor (1915), James H. Jeans (1917), Ralph H. Fowler (1924), Harold Jeffreys (1926), Subrahmanyan Chandrasekhar (1948), George Batchelor (1950), and Abdus Salam (1958).

248   *Einstein's Relativity in Great Britain*

year were Hermann Bondi, William V.D. Hodge, and A. Geoffrey Walker. Penrose's essay, *An Analysis of the Structure of Spacetime*, was the winner, while Hawking's, *Singularities and the Geometry of Space-time*, was awarded an auxiliary Adams Prize at the same time.

Penrose's essay was not, as far I know, published, but circulated in a 138 pages undated mimeograph form, with Penrose signing as a member of the Palmer Physical Laboratory, Princeton University, where he was spending some time, and the Department of Mathematics, Birkbeck College, University of London, his permanent position then.[7] The first lines of the "Abstract" gave a rough idea of the content of the work (Penrose c. 1966):

> "This essay is concerned with techniques for analyzing the structure of space-time and with certain applications. The motivation, in the first instance, for the introduction of the particular techniques used here, springs from a belief that perhaps ultimately the differentiable manifold picture of space-time may have to be abandoned; or, at least, that the null cone and 'causality' structure of space-time should in some way be regarded as essentially more fundamental than its continuity-differentiability structure. The idea, then, is to reformulate coordinate systems and from the methods of general differential geometry to certain concepts which are <u>tied</u> to the existence of the <u>null cones</u> and even to the particular dimensionality and signature of our space-time. Thus the methods which emerge have to be more tailored to relativity physics than are those techniques which may be regarded as presently 'conventional'."

Hawking's essay summarized his work on global properties of general relativity theory, and in particular, developed a series of cosmological singularity theorems he had proved. In the "Abstract" of the essay, Hawking (1966d, 2014: 413) explained that "the definition of a singularity is given in terms of geodesic incompleteness", and that "the various energy assumptions of needed to prove the occurrence of singularities are discussed and then a number of theorems are presented which prove the

---

[7] Penrose's contribution to 1967 Battelle Rencontres, "Structure of Space-Time" (Penrose 1968), includes parts of his Adams Prize essay

The Renaissance of Relativity in Great Britain 249

occurrence of singularities in most cosmological solutions." His conclusion was that "either the General Theory of Relativity breaks down or that there could be particles whose histories did not exist before (or after) a certain time," and he thought that "the theory probably does breaks down only when quantum gravitational effects become important," that is, when the radius of curvature of space-time became about 10^{-14} cm.

Actually, some of the contents of Hawking's essay had been published before, by Hawking himself (1965, 1966b, 1966c), and by Hawking and Ellis (1965). This, together with the priority of Penrose's global techniques helps to understand why it was Penrose who won the "first" prize.

6.3. *The Large-Scale Structure of Space-Time* (1973)

In the years following his Ph.D. dissertation Hawking published several papers dealing with singularities and gravitational collapse, including one together with Robert Penrose (Hawking and Penrose 1970), in which they presented a new theorem on space-time singularities which incorporated previous known results, plus others papers on matters such as gravitational radiation in an expanding universe, the rotation of the universe, but the culmination of his studies on the structure of space-time was the book it published together with George Ellis, *The Large Scale Structure of Space-Time* (Hawking and Ellis 1973). In his Foreword to the 50th anniversary edition, Abhay Ashtekar (2023: x) remarked its importance:

> "In 1923, Cambridge University Press published Arthur Eddington's monograph, *The Mathematical Theory of Relativity*, arguably the first systematic and comprehensive textbook on the theory. It embodies Eddington's view that 'The investigation of the external world is a quest for structure rather than substance'. It had a deep influence on how researchers thought of general relativity in subsequent decades.
>
> Five decades later, the Press published another monograph, *The Large Scale Structure of Space-Time* by Stephen Hawking and George Ellis in 1973. Hailed immediately as 'a masterpiece, written by sure hands' it too focuses on 'structure' — but now on *global aspects* of spacetime structure, which had been almost entirely ignored in earlier books. The monograph solidified the new approach to understand gravitational phenomena, introduced by Roger Penrose through his use of

global methods and causal structures, which transformed the way the community thought of strong gravity. It has even greater impact on the development of relativistic gravity than Eddington's monograph because it helped shape the 'golden age' of general relativity during the 1970s.

Before the appearance of this monograph, contributions to general relativity were by and large dominated by tensor calculus and partial differential equations in local coordinates. The monograph served as a powerful catalyst that changed our way of understanding the physics of general relativity. Thanks in large part to its influence, a sizable fraction of researches started thinking invariantly, in geometrical terms, using spacetime diagrams and light cones. The emphasis shifted to global issues. In subsequent years, this shift led to numerous novel directions that created new frontiers of research."

By then, there were many physicists and mathematicians around the world who were working on the structure of space-time in general relativity, people like Werner Israel, Robert Geroch, Charles W. Misner, Kip S. Thorne or John A. Wheeler, the last three authors of an influential — and big, 1035 pages — book, *Gravitation*, published in 1973, he same year that *The Large Scale Structure of Space-Time*. It was no longer a mainly British field of research, but a world one.

6.4. Towards a Quantum Theory of Gravity: Thermodynamics and Particle Creation in Black Holes

"We were so successful with the classical general theory of relativity that I was a bit of a loose end in 1973, after the publication with George Ellis of *The Large Scale Structure of Spacetime*. My work with Penrose had shown that general relativity broke down at singularities. So the obvious next step would be to combine general relativity, the theory of the very large, with quantum theory, the theory of the very small." This wrote Stephen Hawking (2003: 113) in the collective book celebrating his 60th birthday. Indeed, combining general relativity, a classical theory, with quantum theory was an obvious necessity and felt so. Already in 1918, in a short paper published in *Nature*, Arthur Eddington (1918c: 36) referred

The Renaissance of Relativity in Great Britain   251

to this problem, though as part of his idiosyncratic approach:[8] "From the constant of gravitation, together with the other fundamental constants of Nature — the velocity of light and the quantum of action — it is possible to form a new fundamental unit of length. This unit is 7×10^{-28} cm. It seems to be inevitable that this length must play some fundamental part in any complete interpretation of gravitation. [...] In recent years great progress has been made in knowledge of the excessive minute; but until we can appreciate details of structure down to the quadrillionth or quintillionth of a centimeter, the most sublime of all the forces of Nature remains outside the purview of the theories of physics." Théophile de Donder, Oskar Klein, Vladimir Fock, Hermann Weyl or Erwin Schrödinger were among those who tried to advance in the solution of the problem.[9] "Bryce DeWitt recalls — it was cited in the "Preface" of the book mentioned in the previous footnote — having a conversation with Pauli in 1949, at the Institute for Advanced Study, in which Pauli asked him what he was working on, to which DeWitt responded: 'trying to quantize the gravitational field.' Pauli was something of a veteran of quantum gravity at the time. His response, after shaking and nodding his head a few times ('die Paulibewegung'), was: 'That is a very important problem — but it will take someone really smart!'."[10]

And now back to Stephen Hawking.

Already in 1969, Hawking showed interest in the connection between general relativity-quantum theory, though approaching the problem more from Einstein's gravitational formulation than from the quantum. It was in an essay he submitted to the Gravity Research Foundation Awards founded in 1949 by Roger W. Babson (1875–1967) "to encourage scientific research and to arrive at a more complete understanding of the phenomenon of gravitation [...] with the expectation that beneficial uses will arrive." Hawking essay was entitled "The creation and annihilation of matter by a gravitational field," and its "Summary" read as follows: "It is shown that in classical general relativity, if space-time is nonempty at one time, it will be nonempty at all times provided that the energy-momentum

---

[8] See also his *Relativity Theory of Protons and Electrons* (Eddington 1936).

[9] Their papers, as well as other are reproduced in Blum and Rickles, eds. (2018).

[10] DeWitt was one of the great pioneers of quantum gravity. See DeWitt-Morette (2011).

252 *Einstein's Relativity in Great Britain*

tensor of the matter satisfies a physically reasonable condition. The apparent contradiction with the quantum predictions for the creation and annihilation of matter particles by gravitons is discussed and is shown to arise from the lack of a good energy momentum operator for the matter which obeys the covariant conservation equation." But Hawking did not won the award, occupying the fifth position; the winner was the work presented by V.W. Hughes and W.L. Williams, of the Gibbs Laboratory, Physics Department, Yale University, "Experimental test for mass anisotropy based on nuclear magnetic resonance."

Four years afterwards, Hawking published an important article together with the American physicist (his thesis supervisor had been Richard Feynman) James M. Bardeen and the Australian born and another of Sciama students, Brandon Carter: "The four laws of black hole mechanics" (Bardeen, Carter and Hawking 1973). "Expressions are derived for the mass of a stationary axisymmetric solution of the Einstein equations — read the abstract of this article — containing a black hole surrounded by matter and for the difference in mass between two neighboring such solutions. Two of the quantities which appear in these expressions, namely the area A of the event horizon and the "surface gravity" κ of the black hole, have a close analogy with entropy and temperature respectively. This analogy suggests the formulation of four laws of black hole mechanics which correspond to and in some ways transcend the four laws of thermodynamics." Its main importance lies in the introduction of entropy and temperature in the analysis of black holes."

However, a year before a student of John A. Wheeler in Princeton, Jacob David Bekenstein (1947–2015), had defended a doctoral dissertation entitled *Baryon Number, Entropy and Black Hole Physics* (Bekenstein 1972a).[11] "It has been suggested — Bekenstein (1972a: ix) wrote in the

---

[11] Bekenstein was born in Mexico City from Polish Jews who immigrated to Mexico. He moved early to the United States, gaining U.S. citizenship in 1968. After graduating at the Polytechnic Institute of Brooklyng, New York, he went to Princeton, becoming afterwards a postdoctoral fellow at the University of Texas at Austin from 1972 to 1974, when he immigrated to Israel, where he joined the Ben-Gurion University, becoming there full professor in 1978.

'Abstract' — that the second law of thermodynamics can be circumvented in black hole physics. The argument is that the entropy of the universe can be decreased by the appearance of some of it down a black hole. However, we show that according to some recent evidence, the area of a black hole always increases irreversibly by a non-negligible amount when something goes down the black hole. It thus appears that the decrease of entropy is associated with an irreversible change of the black hole. We suggest that this phenomenon is a manifestation of the second law in generalized form: 'Common entropy plus black-hole entropy never decreases.' We then arrive at a definition of the entropy of a black hole as a function of its area in terms of which he generalized second law holds even when common entropy disappears down a black hole." That same year, Bekensten (1972b) presented part of his Ph.D. results in a short paper in *Lettere al Nuovo Cimento*: "Black holes and the second law". The following year, Bekenstein (1973) made his results more widely known through an article published in *Physical Review*: "Black holes and entropy."

And so, the field of black hole physics and thermodynamic became united. Of such development, Dennis Sciama (1976: 385) would write:

> "The last few years have seen a dramatic increase in our theoretical understanding of black holes and in our observational prospects of detecting them. A large part of the theoretical advances are confined to the theory of gravitation itself, that is, to Einstein's general theory of relativity. However, among these results, a particular set stands out by virtue of possessing a strong connection with the rest of physics. These results involve the thermodynamic properties of black holes. Since thermodynamics and essentially universal character in non-gravitational physics, it is of great scientific importance that its scope is wide enough to apply also to processes involving black holes."

In 1974, Stephen Hawking announced a new result, of great importance, using what can be termed a "semiclassical approach" to the quantization of gravity. He presented that result during a "Quantum Gravity Symposium" held at the Rutherford Laboratory, Chilton, Oxford, on 15–16 February 1974. Its title was "Particle creation by black holes," an obvious reminiscence of his Gravity Research Foundation Awards essay.

254   *Einstein's Relativity in Great Britain*

Hawking's paper began as follows (Hawking 1975a: 219):

"Although there has been a lot of work in the last fifteen years, I think it would be fair to say that we do not yet have a fully satisfactory and consistent quantum theory of gravity. At the moment classical general relativity still provides the most successful description of gravity. In classical general relativity one has a classical metric which obeys the Einstein equations, the right-hand side of which is supposed to be the energy-momentum tensor of the classical metric fields. However, although it may be reasonable to ignore quantum gravitational effects on the grounds that these are likely to be small, we know that quantum mechanics plays a vital role in the behavior of the matter fields. One therefore has the problem of defining a consistent scheme in which the space-time metric is treated classically but is coupled to the matter fields which are treated quantum mechanically. Presumably such a scheme would be only an approximation to a deeper theory (still to be found) in which space-time itself was quantized. However one would hope that it would be a very good approximation for most purposes except near space-time singularities."

The nucleus of Hawking's (1975a: 223–224) idea was that "when the radius of curvature of space-time is smaller than the Compton wavelength of a given species of particles one gets an indeterminacy in the particle number, or, in other words, particle creation", and "even though the effects of particle creation may be negligible locally [...] they can add up to have a significant influence on black holes over the lifetime of the universe [...]". Therefore,

"the gravitational field of a black hole will create particles and emit them to infinity at just the rate that one would expect if the black hole were an ordinary body with a temperature in geometric units of $k/2\pi$ where k is the 'surface gravity' of the black hole. In ordinary units this temperature is of the order of $10^{26}M^{-1}$ ^{0}K, where M is the mass, in grammes, of the black hole. For a black hole of solar mass (10^{33} g) this temperature is much lower than the 3 ^{0}K temperature of the cosmic microwave background. Thus black holes of this size would be absorbing radiation faster

The Renaissance of Relativity in Great Britain 255

than they emitted it and would be increasing in mass. However, in addition to black holes formed by stellar collapses, there might also be much smaller black holes which were formed by density fluctuations in the very early universe. These small black holes, being at a higher temperature would radiate more than they absorbed. They would therefore presumably decrease in mass."

In principle, the papers presented at the Oxford Symposium were not intended to be published, and although they were so finally, the diffusion of a book was much lower than an article. Thus, Hawking published also his result in the German journal — edited by Springer — *Communications in Mathematical Physics* (Hawking 1975b).

In the above cited article, Sciama (1976: 385) emphasized the importance of Hawking's result:

"it was from calculations based on first principles that Stephen Hawking discovered in 1974 his formula for the quantum radiance of a black hole. According to Hawking, this radiance has a thermal character; the radiation temperature of a black hole being precisely the quantity that previously had been called 'analogous' to the temperature of a black hole. If this discovery is confirmed by more complete calculations, it will represent one of the most important theoretical advances ever made in physics. In particular, it would enable one to complete the chain of argument which establishes the thermodynamics behavior of stationary black holes, so linking them to the rest of physics."

Hawking thought that he had found out a very important result. Representative of what he thought is what he wrote in his "Sixty years in a nutshell" essay, published as we saw in a collective volume celebrating his 60th birthday (Hawking 2003: 113):

"I studied how quantum fields would scatter off a black hole. I was expecting that part of an incident wave would be absorbed and the remainder scattered. But to my great surprise, I found there seemed to be emission from the black hole. At first, I thought this must be a mistake in my calculation. But what persuaded me that it was real, was that the emission was exactly what was required to identify the area of the horizon, with the entropy of a black hole,

$$S = \frac{c^3 kA}{4\hbar G}$$

I would like this simple formula to be on my tombstone."

And indeed, following his wish his ashes at Westminster Abbey are laid under a stone in which that formula appears.

However, today that expression for the entropy of a black hole is named not only after Hawking but after Bekenstein also: the "Bekenstein-Hawking entropy formula."

As it turns out, Hawking was critical of Bekenstein's previous work. In "Particle creation by black holes," he wrote (Hawking 1975a: 228; 1975b): "Bekenstein [1973] suggested that A and k were not merely analogous to entropy and temperature respectively but hat, in some sense, they actually were the entropy and temperature of the black hole. Although the ordinary second law of thermodynamics is transcended in that entropy can be lost down black holes, the flow of entropy across the event horizon would always cause some increase in the area of the horizon. Bekenstein [1974] therefore suggested a Generalized Second Law: Entropy + some multiple (unspecified) of A never decreases. However, he did not suggest that a black

The Renaissance of Relativity in Great Britain 257

hole could emit particles as well as absorb them. Without such emission the Generalized Second Law would be violated by, for example, a black hole immersed in black body radiation at a lower temperature than that of the black hole. On the other hand, if one accepts that black holes do emit particles at a steady rate, the identification of k/2π with temperature ¼·A with entropy is established and a Generalized Second Law confirmed."

From this quotation, it seems obvious that Hawking was not only critical but somewhat irritated with Bekenstein's work. Such impression was explicated by Eliezer Rabionovici (2020: xii) in a volume dedicated to the memory of Bekenstein:

> "Physics may be about uncovering objective secrets of nature, but it is done by humans, and most humans appreciate due recognition and credit. Jacob's [Bekenstein] ideas in the 1970s encountered both significant resistance not least from Stephen Hawking, who briefly mentions Bekenstein's role in his bestselling book *A Brief History of Time*. He writes: 'I was motivated partly by irritation with Bekenstein, who I felt misused my discovery.' And then: 'However, it turned out in the end that he [Bekenstein] was basically correct, though in a manner he had certainly not anticipated.'"

Indeed, physics is done by humans, but contrary to many other human activities, finally, or more often than not, the right credits are assigned. So it was in this case, though the merit of Hawking's idea of black hole radiance was an important step in the course of developing a quantum theory of gravity, still to be obtained.

In some sense, the confrontation Hawking–Bekenstein shows that the research in relativity, in general relativity, was a truly international endeavor. Though it is true that the Department of Applied Mathematics and Theoretical Physics, of Cambridge University, where Stephen Hawking was professor (indeed in 1979 he became Lucasian Professor, the chair once held by Isaac Newton, as well as Paul Dirac from 1932 to 1969), and the Mathematical Institute of Oxford University, the academic home of Roger Penrose, where he was Rouse Ball Professor of Mathematics, were internationally renowned centers of relativity studies, Einstein's theory was no longer a, above all, German or British field. Therefore, it is right to stop this book here.

References

Abir-Am, P. G. (1987). "Synergy or clash: Disciplinary and marital strategies in the career of mathematical biologist Dorothy Wrinch," in Abir-Am and Outram, eds. (1987: 239–280).

Abir-Am, P. G. and Outram, Dorinda, eds. (1987). *Uneasy Careers and Intimate Lives. Women in Science, 1789–1979* (Rutgers University Press, New Brunswick).

Abraham, M. (1905). *Theorie der Elektrizitiit: Elektromagnetische Theorie der Strahlung* (Teubner, Leipzig).

Abraham, M. (1913). "Eine neue Gravitationstheorie," *Archiv der Mathematik und Physik 20*, 193–209. English translation in Renn and Schemmel, eds (2007: 347–362).

Alexander, S. (1920). *Space, Time and Deity: The Glifford Lectures at Glasgow, 1916–1918* (Macmillan, London).

Alter, P. (1987). *The Reluctant Patron of Science and the State in Britain 1850–1920* (Berg, Oxford).

Abhay, A. (2023). "Foreword," in 50th anniversary edition (2023) of Stephen W. Hawking and George F. R. Ellis, *The Large Scale Structure of Space-Time* (1973). pp. x–xii.

Ayer, A. J. (1987). *Philosophy in the Twentieth Century* (Unwin Paperbaks, London; first edition of 1982).

Baldwin, O. R. (1929). "The Relativity Theory of divergent waves," *Proceedings of the Royal Society of London A 123*, 119–133.

Baldwin, O. R. and Jeffery, G. B. (1926). "The Relativity Theory of plane waves," *Proceedings of the Royal Society of London A 111*, 95–104.

Bardeen, J. M., Carter, B. and Hawking, S. W. (1973). "The four laws of black hole mechanics," *Communications in Mathematical Physics 31*, 161–170.

260 *Einstein's Relativity in Great Britain*

Bateman, H. (1909). "The Conformal Transformations of a Space of Four Dimensions and Their Applications to Geometrical Optics," *Proceedings of the London Mathematical Society 7*, 70–89.

Bateman, H. (1910). "The Relation between Electromagnetism and Geometry", *Philosophical Magazine 19*, 623–628.

Bateman, H. (1915). *The Mathematical Analysis of Electrical and Optical Wave-Motion* (Cambridge University Press, Cambridge).

Becquerel, J. (1922). *Le principle de relativité et la théorie de la gravitation. Leçons profesées en 1921 et 1922 a l'Ecole Polytechnique et au Muséum d'Histoire Naturelle* (Gauthier-Villars, Paris).

Bekenstein, J. D. (1972a). *Baryon Number, Entropy, and Black Hole Physics*, dissertation presented to the Faculty of Princeton University in candidacy for the degree of Doctor of Philosophy (Princeton University).

Bekenstein, J. D. (1972b). "Black holes and the Second Law," *Lettere al Nuovo Cimento 4*, 737–740.

Bekenstein, J. D. (1973). "Black holes and entropy," *Physical Review D 7*, 2333–2346.

Bekenstein, J. D. (1974). "Generalized second law of thermodynamics in black hole physics," *Physical Review D 9*, 3292–3300.

Bergia, S. (1993). "Attemps at unified field theories (1919–1955). Alleged failures and intrinsic validation/refutation criteria," in Earman, Janssen and Norton, eds. (1993: 274–307).

Bergson, H. (1922). *Durée et simultanéité. A propos de la théorie d'Einstein* (Librairie Félix Alcan, París).

Bergmann, P. G. and Komar, A. B. (1962). "Observables and commutation relations observables et regles de commutation." In Royaumont (1962: 309–323).

Bertotti, B., Balbinot, R., Bergia, S. and Messina, A., eds. (1990). *Modern Cosmology in Retrospect* (Cambridge University Press, Cambridge).

Biezunski, M. (1981). *La diffusion de la théorie de la relativité en France.* Doctoral dissertation, University of Paris.

Biezunski, M., ed., (1989). *Albert Einstein. Oeuvres choisies*, vol. 4 (*Correspondances françaises*) (Éditions du Seuil, Paris).

Blum, A. S. and Rickles, D., eds. (2018). *Quantum Gravity in the First Half of the Twentieth Century* (Edition Open Sources, Max Planck Institute for the History of Science, Berlin).

Bondi, H. (1947). "Spherically symmetric models in General Relativity," *Monthly Notices of the Royal Astronomical Society 107*, 420–425.

Bondi, H. (1952). *Cosmology* (Cambridge University Press, Cambridge).

Bondi, H. (1956). "The steady-state theory of cosmology and relativity," in *Fünfzig Jahre Relativitätstheorie — Cinquantenaire de la Théorie de la Relativité. Jubilee of Relativity Theory*, Mercier, André, and Kervaire, Michel, eds., *Helvetica Physical Acta. Supplementum IV* (Birkhäuser Verlag, Basel), pp. 152–154.

Bondi, H. (1969). "Gravitational bounce in General Relativity," *Monthly Notices of the Royal Astronomical Society 142*, 333–353.

Bondi, H. (2004). "Gravitational waves in General Relativity. XVI. Standing waves," *Proceedings of the Royal Society A 460*, 463–470.

Bondi, H. and Gold, T. (1948). "The Steady-State theory of the expanding universe," *Monthly Notices of the Royal Astronomical Society 108*, 252–270.

Bondi, H. and McCrea, W. H. (1960). "Energy transfer by gravitation in Newtonian theory," *Proceedings of the Cambridge Philosophical Society 56*, 410–413.

Bondi, H., Pirani, F. A. E. and Robinson, I. (1959). "Gravitational waves in general relativity III. Exact plane waves," *Proceedings of the Royal Society A 251*, 519–533.

Bondi, H., Van der Burg, M. G. J. and Metzner, A. W. K. (1963). "Gravitational waves in general relativity VII. Waves from axi-symmetric isolated systems," *Proceedings of the Royal Society A 269*, 21–52.

Bridgman, P. W. (1927). *The Logic of Modern Physics* (Macmillan, New York).

Briginshaw, A. J. (1979). "The axiomatic geometry of Space-Time: An assessment of the work of A. A. Robb," *Centaurus 22*, 315–323.

Broad, C. D. (1915). "What do we mean by the question: is our space Euclidean?," *Mind 24*, 464–480.

Broad, C. D. (1916). "The nature and geometry of space," *Mind 25*, 522–524.

Broad, C. D. (1920). "Critical notices: *The Principles of Natural Philosophy*. By A. N. Whitehead," *Mind 29*, 216–231.

Broad, C. D. (1921a). "Review of E. Freundlich, *The Foundations of Einstein's Theory of Gravitation*," *Mind 30*, 101–102.

Broad, C. D. (1921b). "Review of M. Schlick, *Space and Time in Contemporary Physics*," *Mind 30*, 245.

Broad, C. D. (1921c). "Review of E. Cunningham, *Relativity, the Electron Theory and Gravitation*," *Mind 30*, 490.

Broad, C. D. (1921d). "Review of A. A. Robb, *The Absolute Relations of Space and Time*," *Mind 30*, 490.

Broad, C. D. (1923a). *Scientific Thought* (Kegan Paul, London).

262 *Einstein's Relativity in Great Britain*

Broad, C. D. (1923b). "Critical notices. *The Principle of Relativity, with Applications to Physical Science*. By A. N. Whitehead," *Mind 32*, 211–219.

Broad, C. D. (1925). "*La déduction relativiste*. By Émile Meyerson," *Mind 34*, 504–505.

Broad, C. D. (1948). "Alfred North Whitehead (1861–1947)," *Mind 57*, 139–145.

Bucherer, A. H. (1907). "On a new principle of Relativity on electromagnetism," *Philosophical Magazine 13*, 413–421.

Buchwald, J. Z. (1981a). "The abandonment of Maxwellian Electrodynamics: Joseph Larmor's Theory of the Electron." *Archives Internationales d'Histoire des Sciences 31*, 135–180.

Buchwald, J. Z. (1981b). "The abandonment of Maxwellian Electrodynamics: Joseph Larmor's Theory of the Electron. Part II." *Archives Intemationales d'Histoire des Sciences 31*, 373–438.

Buchwald, J. Z. (1985). *From: Maxwell to Microphysics* (The University of Chicago Press, Chicago).

Campbell, J. E. (1903). *Introductory Treatise on Lies Theory of Finite Continuous Transformation Groups* (Clarendon Press, Oxford); reprinted by Chelsea (New York) in 1966.

Campbell, J. E. (1920). "Einstein's theory of gravitation as an hypothesis in differential geometry," *Proceedings of the London Mathematical Society 20*, 1–14.

Campbell, J. E. (1922). "On a class of surfaces in Euclidean space which generate an expression for the space time interval in Einstein's geometry of a particular form," *Proceedings of the London Mathematical Society 21*, 317–324.

Campbell, J. E. (1923). "On the generation of the ground form of the statical gravitation field in Einstein's theory of gravitation from the ground form of any surface in ordinary space," *Proceedings of the London Mathematical Society 22*, 92–103.

Campbell, J. E. (1926). *A Course of Differential Geometry* (Clarendon Press, Oxford).

Campbell, N. R. (1907). *Modern Electrical Theory* (Cambridge University Press, Cambridge).

Campbell, N. R. (1910). "The aether," *Philosophical Magazine 19*, 129–137.

Campbell, N. R. (1913). *Modern Electrical Theory* (second edition, Cambridge University Press, Cambridge).

Campbell, N. R. (1921). *What is Science?* (Methuen, London).

Campbell, N. R. (1923). *Modem Electrical Theory. Supplementary Chapters* (Cambridge University Press, Cambridge).

Cantor, G. N. and Hodge, M. J. S., eds. (1981). *Conceptions of Ether* (Cambridge University Press, Cambridge).

Carnap, R. (1921). *Der Raum. Ein Beitrag zur Wissenschaftslehre* (University of Jena, Jena).

Carnap, R. (1963). "Intellectual autobiography," in Schilpp, ed. (1963: 3–84).

Carr, H. W. (1920). *The General Principle of Relativity, in its Philosophical and Historical Aspect* (Macmillan and Co., London).

Carr, H. W. (1921). "Review of *The Reign of Relativity*," *Mind 30*, 462–467.

Carr, H. W. (1922). "Einstein's theory and philosophy," *Mind 31*, 169–177.

Carr, H. W., Nunn, T. P., Whitehead, A. N. and Wrinch, D. (1922). "Discussion: The idealistic interpretation of Einstein's theory," *Proceedings of the Aristotelian Society 22*, 123–138.

Cassirer, E. (1953). *Substance and Function & Einstein's Theory of Relativity* (Dover, New York).

Chandrasekhar, S. (1983). *Eddington. The most distinguished astrophysicist of his time* (Cambridge University Press, Cambridge).

Clark, G. L. (1954). "The problem of two bodies in Whitehead's theory," *Proceedings of the Royal Society of Edinburgh A 64*, 49–56.

Combridge, J. T. (1965). *Bibliography of Relativity and Gravitation Theory 1921 to 1937* (King's College, London).

Conway, A. W. (1936–1938). "Professor G. A. Schott, 1868–1937," *Obituary Notices of Fellows of the Royal Society 2*, 451–454.

Copson, E. T. (1928). "On electrostatics in a gravitational Field," *Proceedings of the Royal Society of London A 118*, 184–194.

CPAE. (1993). *The Collected Papers of Albert Einstein*, vol. 5 (*"The Swiss Years. Correspondence, 1902–1914"*), Klein, M. J., Kox, A. J. and Schulmann, R., eds. (Princeton University Press, Princeton).

CPAE. (2004). *The Collected Papers of Albert Einstein*, vol. 9 (*"The Berlin Years: Correspondence, January 1919–April 1920"*), Kormos Buchwald, Diana, Schulmann, Robert, Illy, József, Kennefick, Daniel J. y Sauer, Tilman, eds. (Princeton University Press, Princeton).

CPAE. (2006). *The Collected Papers of Albert Einstein*, vol. 10 (*The Berlin Years: Correspondence, May–December 1920, and Supplementary Correspondence, 1919–1920*, Kormos Buchwald, Diana, Sauer, Tilman, Rosenkranz, Ze'ev, Illy, József, and Holmes, Virginia I., eds. (Princeton University Press, Princeton).

CPAE. (2021). *The Collected Papers of Albert Einstein*, vol. 16 (*The Berlin Years: Writings & Correspondence, June 1927–May 1929*, Kormos Buchwald, Diana, Ze'ev Rosenkranz, Ze'ev, Illy, József, Kennefick, Daniel J., Kox, A.

264 *Einstein's Relativity in Great Britain*

J., Lehmkhhl, Dennis, Sauer, Tilman, and Nollan James, Jennifer, eds. (Princeton University Press, Princeton).

Crelinsten, J. (2006). *Einstein's Jury. The Race to Test Relativity* (Princeton University Press, Princeton).

Crommelin, A. C. D. (1931). "The President's Address," *Monthly Notices* of the *Royal Astronomical Society 91*, 422–434.

Cunningham, E. (1907). "On the electromagnetic mass of a moving electron," *Philosophical Magazine 14*, 538–547.

Cunningham, E. (1910). "The Principle of Relativity in Electrodynamics and an extension thereof," *Proceedings of the London Mathematical Society 8*, 77–98.

Cunningham, E. (1914). *The Principle of Relativity* (Cambridge University Press, Cambridge).

Cunningham, E. (1921). *Relativity, the Electron Theory, and Gravitation* (Longmans & Green, New York).

Curzon, H. E. J. (1924a). "Bipolar solutions of Einstein's gravitation equations," *Proceedings of the London Mathematical Society 23*, xxix.

Curzon, H. E. J. (1924b). "Cylindrical solutions of Einstein's gravitation equations," *Proceedings of the London Mathematical Society 23*, 477–480.

De Sitter, W. (1916a). "On Einstein's theory of gravitation and its astronomical consequences," *Monthly Notices* of the *Royal Astronomical Society 76*, 699–728.

De Sitter, W. (1916b). "On Einstein's theory of gravitation and its astronomical consequences. Second Paper," *Monthly Notices of the Royal Astronomical Society 77*, 155–184, 481.

De Sitter, W. (1917). "On Einstein's theory of gravitation and Its astronomical consequences. Third Paper," *Royal Astronomical Society. Monthly Notices 78*, 3–28.

De Swart, J. (2020). "Closing in on the Cosmos: Cosmology's rebirth and the rise of the dark matter problem," in *The Renaissance of General Relaivity in Context*, Blum, Alexander S., Lalli, Roberto, and Renn, Jürgen, eds. (Birkäuser, Cham), pp. 257–284.

DeVorkin, D., ed. (1999). *The American Astronomical Society's First Century* (American Astronomical Society, Washington D. C.).

DeVorkin, D. (2000). *Henry Norris Russell. Dean of American Astronomers* (Princeton University Press, Princeton).

DeWitt, C. M., ed. (1957). *Conference on the Role of Gravitation in Physics*, WADC Technical Report 57–216, ASTIA Document No. AD 118189 (Wright Air Development Center, Ohio).

DeWitt-Morette, C. (2011). *The Pursuit of Quantum Gravity. Memoirs of Bryce DeWitt from 1946 to 2004* (Springer, Berlin).

Dingle, H. (1923). *Relativity and Gravitation* (Methuen, London).

Dirac, P. A. M. (1924a). "Dissociation under a temperature gradient," *Proceedings of the Cambridge Philosophical Society 22*, 132–137.

Dirac, P. A. M. (1924b). "Note on the relativity dynamics of a particle," *Philosophical Magazine 47*, 1158–1159.

Dirac, P. A. M. (1928). "The quantum theory of the electron." *Proceedings of the Royal Society* (London) *A 117*, 610–624.

Dirac, P. A. M. (1975). *General Theory of Relativity* (Wiley-Interscience, New York).

Dirac, P. A. M. (1982). "The early years of relativity," in Holton and Elkana, eds. (1982: 79–90).

Donnan, F. G. (1919). "Heat of reaction and gravitational field," *Nature 104*, 392–393.

Douglas, A. V. (1956). *The Life of Arthur Stanley Eddington* (Thomas Nelson & Sons, London).

Dreyer, J. L. E. and Turner, H. M., eds. (1923). *History of the Royal Astronomical Society, 1820–1920* (Royal Astronomical Society, London. Reprinted by Blackwell, Oxford 1987).

Duffield, W. G. (1919–1920). "The displacement of spectrum lines and the Equivalence Hypothesis." *Monthly Notices of the Royal Astronomical Society 80*, 262–272.

Dyson, F. W. (1917). "On the opportunity afforded by the eclipse of 1919 May 29 of verifying Einstein's theory of gravitation," *Monthly Notices of the Royal Astronomical Society 77*, 445–447.

Dyson, F. W., Eddington, A. S. and Davidson, C. (1920). "A determination of the deflection of light by the Sun's gravitational field," *Philosophical Transactions of the Royal Society of London A 220*, 291–333.

Earman, J. and Glymour, C. (1980). "Relativity and eclipses: The British eclipse expeditions of 1919 and their predecessors," *Historical Studies in the Physical Sciences 11*, 49–85.

Earman, J., Janssen, M. and Norton, J. D. eds. (1993). *The Attraction of Gravitation* (Birkhäuser, Boston).

Eckert, Michael (1996). "Theoretical physicists at war: Sommerfeld students in Germany and as emigrants," in Forman and Sánchez-Ron, eds. (1996: 69–86).

Eddington, A. S. (1914). *Stellar Movements and the Structure of the Universe* (Macmillan, London).

266 *Einstein's Relativity in Great Britain*

Eddington, A. S. (1915). "Gravitation," *Observatory 38*, 93–98.

Eddington, A. S. (1916). "Gravitation and the principle of relativity." *Nature 98*, 328–331.

Eddington, A. S. (1917a). "Einstein's theory of gravitation," *Monthly Notices Royal Astronomical Society 77*, 377–382.

Eddington, A. S. (1917b). "Astronomical consequences of the electrical theory of matter. Note on Sir Oliver Lodge's suggestions," *Philosophical Magazine 34*, 163–167.

Eddington, A. S. (1917c). "Astronomical consequences of the electrical theory of matter. Note on Sir Oliver Lodge's suggestions, 11," *Philosophical Magazine 34*, 321–327.

Eddington, A. S. (1918a). "Electrical theories of matter and their astronomical consequences with special reference to the Principle of Relativity," *Philosophical Magazine 35*, 481–487.

Eddington, A. S. (1918b). *Report on the Relativity Theory of Gravitation* (Fleetway Press-The Physical Society of London, London).

Eddington, A. S. (1918c). "Gravitation and the Principle of Relativity II," *Nature 101*, 34–36.

Eddington, A. S. (1920a). "Discussion on Einstein's theory of relativity," March 26, 1920, *Proceedings of the Physical Society of London 32*, 245–251.

Eddington, A. S. (1920b). *Report on the Relativity Theory of Gravitation*, second edition (Fleetway Press-The Physical Society of London).

Eddington, A. S. (1920c). *Space, Time and Gravitation: An Outline of the General Theory of Relativity* (Cambridge University Press, Cambridge).

Eddington, A. S. (1920d). "The meaning of matter and the laws of nature according to the theory of relativity," *Mind 29*, 145–158.

Eddington, A. S. (1921a). *Espace, temps et gravitation* (Hermann, Paris).

Eddington, A. S. (1921b). "A generalization of Weyl's theory of the electromagnetic and gravitational fields," *Proceedings of the Royal Society A 99*, 104–122.

Eddington, A. S. (1922). "The Propagation of gravitational waves," *Proceedings of the Royal Society of London A 102*, 268–282.

Eddington, A. S. (1923). *The Mathematical Theory of Relativity* (Cambridge University Press, Cambridge).

Eddington, A. S. (1924). "A comparison of Whitehead's and Einstein's formulae," *Nature 113*, 192.

Eddington, A. S. (1928). *The Nature of the Physical World* (Cambridge University Press, Cambridge).

Eddington, A. S. (1930a). "On the instability of Einstein's spherical world," *Monthly Notices of the Royal Astronomical Society 90*, 668–678.

Eddington, A. S. (1930b). "Space and its properties," *Nature 125*, 849–850.

Eddington, A. S. (1931). "The expansion of the universe (Council note)," *Monthly Notices of the Royal Astronomical Society 91*, 412–416.

Eddington, A. S. (1932). "The expanding universe," *Proceedings Physical Society 44*, 1–16.

Eddington, A. S. (1936). *Relativity Theory of Protons and Electrons* (Cambridge University Press, Cambridge).

Eddington, A. S. (1939). *The Philosophy of Physical Science* (Cambridge University Press, Cambridge).

Eddington, A. S. (1942–1944). "Joseph Larmor, 1857–1942," *Obituary Notices of Fellows of the Royal Society 4*, 197–207.

Eddington, A. S. (1943). *The Combination of Relativity Theory and Quantum Theory* (The Dublin Institute for Advanced Studies, Dublin).

Eddington, A. S. (1953). *Fundamental Theory* (Cambridge University Press, Cambridge).

Eddington, A. S. and Clark, G. L. (1938). "The problem of n bodies in General Relativity Theory," *Proceedings of the Royal Society of London A 166*, 465–475.

Eddington, A. S. and Cottingham, E. T. (1920). "Photographs taken at Príncipe during the total eclipse of the Sun, May 29th," In *Report of the Eighty-Seventh Meeting of the British Association for the Advancement of Science* (John Murray, London), pp. 156–157.

Eddington, A. S., Ross, W. D., Broad, C. D. and Lindemann, F. A. (1920). "The philosophical aspect of the theory of relativity," *Mind 29*, 415–445.

Einstein, A. (1928). "Riemann-Geometrie mit Aufrechterhaltung des Begriffes des Fern-parallelismus," *Preussische Akademie der Wissenschaften. Phys.-mathe. Klasse, Sitzungsberichte*, pp. 217–221.

Einstein, A. (1931). "Zum kosmologischen Problem der allgemeinen Relativitätstheorie," *Preussische Akademie der Wissenschaften. Phys.-mathe. Klasse, Sitzungsberichte*, pp. 235–237.

Einstein, A. and de Sitter, W. (1932). "On the relation between the expansion and the mean density of the universe," *Proceedings of the National Academy of Sciences 18*, 213–214.

Einstein, A. and Fokker, A. D. (1914). "Nordstromsche Gravitations theorie vom Standpunkt des absoluten Differentialkalküls." *Annalen der Physik 44*, 321–328.

268   *Einstein's Relativity in Great Britain*

Einsenstaedt, J. (1993). "Lemaître and the Schwarzschild solution," in Earman, Janssen and Norton, eds. (1993: 353–389).

Einsenstaedt, J. (2003). *Einstein et la relativite générale. Les Chemins de l'espace-temps* (CNRS Editions, Paris).

Eisenstaedt, J. and Kox, A. J., eds. (1992). *Studies in the History of General Relativity* (Birkhäuser, Boston).

Ellis, G. F. R. (2014). "Stephen Hawking's 1966 Adams Prize Essay," *The European Physical Journal H 39*, 403–411.

Ellis, G. F. R. and Penrose, R. (2010). "Dennis William Sciama. 18 November 1926–19 December 1999," *Biographical Memoirs of Fellows of the Royal Society 56*, 401–422.

Eve, A. S. (1939). *Rutherford. Being the Life and Letters of the Rt. Hon. Lord Rutherford, O. M.* (MacMillan, New York).

Evershed, J. (1921). "The Relativity shift in the solar spectrum," *Observatory 44*, 243–245.

Evershed, J. (1928). "The solar rotation and the Einstein displacement derived from measures of the H and K lines in prominences," *Royal Astronomical Society. Monthly Notices 88*, 126–134.

Evershed, J. (1931). "The shift towards red on the Calcium, Aluminium, and Iron lines in the solar spectrum," *Royal Astronomical Society. Monthly Notices 91*, 260–270.

Feigl, H. (1938/39). "Moritz Schlick," *Erkenntnis 7*, 393–419. English version in Feigl (1979).

Feigl, H. (1979). "Moritz Schlick, a memoir," in Schlick (1979a: xv–xxxviii).

Feynman, M., ed. (2005). *Perfectly Reasonable Deviations from the Beaten Track. The Letters of Richard Feynman* (Basic Books, New York).

FitzGerald, G. (1889a). *Report of the Fifty-Eight Meeting of the British Association for the Advancement of Science held at Bath in September 1888* (John Murray, London), pp. 557–562.

FitzGerald, G. (1889b). "The ether and the Earth's atmosphere," *Science 13*, 390.

FitzGerald, G. (1902). *The Scientific Writings of the Late George Francis FitzGerald*, Joseph Larmor, ed. (Hodges, Figgis, & Co, Dublin, and Longmans, Green, and Co, London).

Fokker, A. D. (1915). "A summary of Einstein and Grossmann's theory of gravitation," *Philosophical Magazine 29*, 77–96.

Forman, P. and Sánchez-Ron, J. M., eds. (1996). *National Military Establishments and the Advance of Science and Technology* (Kluwer, Dordrecht).

Forsyth, A. R. (1890–1906). *Theory of Differential Equations*, 6 vols. (Cambridge University Press, Cambridge).

Forsyth, A. R. (1912). *Lectures on the Differential Geometry of Curves and Surfaces* (Cambridge University Press, Cambridge).

Forsyth, A. R. (1920). "Note on the central differential equation in the Relativity Theory of gravitation," *Royal Society of London. Proceedings A 97*, 145–151.

Forsyth, A. R. (1921). "Note on the path of a ray of light in the Einstein Relativity Theory of gravitational effect," *Royal Astronomical Society. Monthly Notices 82*, 2–11.

Gale, G. and Urani, J., "Milne, Bondi and the 'Second Way' to cosmology," in Goenner, Renn and Sauer, eds. (1999: 343–375).

Glick, T. F., ed. (1987a). *The Comparative Reception of Relativity* (Reidel, Dordrecht).

Glick, T. F. (1987b). "Cultural Issues in the Reception of Relativity," in Glick, ed. (1987 a: 381–400).

Godart, O. (1992). "Contributions of Lemaître to general relativity (1922–1934)," in Eisenstaedt and Kox, eds. (1992: 437–452).

Goenner, H. R. J. and Sauer, T., eds. (1999). *The Expanding Worlds of General Relativity* (Birhäuser, Boston).

Goldberg, J. N. (1962). "Conservation laws and equations of motion," in Royaumont (1962: 31–43).

Goldberg, S. (1970). "In defense of ether: The British response to Einstein's theory of relativity, 1904–1911," *Historical Studies in the Physical Sciences 2*, 89–124.

Goldberg, S. (1984). *Understanding Relativity* (Birkhauser, Boston).

Graham, L. R. (1972). *Science and Philosophy in the Soviet Union* (Alfred A. Knopf, New York).

Gray, A. (1920). "Presidential Address," in *Report of the Eighty-Seventh Meeting of the British Association for the Advancement of Science, Bournemouth 1919* (John Murray, London), pp. 135–146.

Greenwood, T. (1920). "Einstein and idealism," *Mind 31*, 205–207.

Guichelaar, J. (2018). *Willen de Sitter. Einstein's Friend and Opponent* (Springer, Cham).

Guide (1994). *Guide to the Archival Collections in the Niels Bohr Library at the American Institute of Physics, International Catalog of Sources for History of Physics and Allied Sciences, report No. 7* (American Institute of Physics, College Park, MD).

270 *Einstein's Relativity in Great Britain*

Haldane, V. (1921). *The Reign of Relativity* (John Murray, London).

Haldane, V. (1929). *An Autobiography* (Hodder and Stoughton Limited, London).

Havas, P. (1989). "The early history of the 'Problem of motion' in General Relativity," in *Einstein and the History of General Relativity*, Howard, Don, and Stachel, John, eds. (Birkhäuser, Boston), pp. 234–276.

Hawking, S. (1965). "Occurrence of singularities in open universes," *Physical Review Letters 15*, 689–690.

Hawking, S. (1966a). *Properties of Expanding Universes*. Thesis presented for the degree of Ph.D. in the University of Cambridge.

Hawking, S. (1966b). "Singularities in the universe," *Physical Review Letters 17*, 444–445.

Hawking, S. (1966c). "The occurrence of singularities in cosmology," *Proceedings of the Royal Society of London A 294*, 511–521.

Hawking, S. (1966d). *Singularities and the geometry of space-time*. Adams Prize essay. Reproduced in Hawking (2014).

Hawking, S. (1975a). "Particle creation by black holes," in *Quantum Gravity. An Oxford Symposium*, Isham, C. J., Penrose, R. and Sciama, D. W., eds. (Claredon Press, Oxford), pp. 219–267.

Hawking, S. (1975b). "Particle creation by black holes," *Communications in Mathematical Physics 43*, 199–220.

Hawking, S. (2003). "Sixty years in a nutshell," in *The Future of Theoretical Physics and Cosmology. Celebrating Stephen Hawking's 60th Birthday* Gibbons, G. W., Shellard, E. P. S. and Rankin, S. J., eds. (Cambridge University Press, Cambridge), pp. 105–117.

Hawking, S. (2013). *My Brief History* (Bantam Books, New York).

Hawking, S. (2014). "Singularities and the geometry of spacetime," *The European Physical Journal H 39*, 413–503.

Hawking S. W. and Ellis, G. F. R. (1965). "Singularities in homogeneous world models," *Physical Review Letters 17*, 246–247.

Hawking, S. W. and Ellis, G. F. R. (1973). *The Large Scale Structure of Space-Time* (Cambridge University Press, Cambridge).

Hawking, S. W. and Penrose, R. (1970). "The singularities of gravitational collapse and cosmology," *Proceedings of the Royal society of London A 314*, 529–548 (1970).

Heckmann, O. (1962). "General review of cosmological theories," in McVittie, ed. (1962: 429–439).

Hentschel, A. (transl.). (2004). *The Collected Papers of Albert Einstein*, vol. 10 (*The Berlin Years: Correspondence, January 1919–April 1920* (Princeton University Press, Princeton).

Hentschel, A. (transl.). (2006). *The Collected Papers of Albert Einstein*, vol. 10 (*The Berlin Years: Correspondence, May–December 1920, and Supplementary Correspondence, 1909–1920* (Princeton University Press, Princeton).

Hentschel, K. (1990). *Interpretationen und Fehlinterpretationen der speziellen und der allgemeinen Relativitätstheorie durch Zeitgenossen Albert Einsteins* (Birkhäuser Verlag, Basel).

Hetherington, N. S. (1993). *Encyclopedia of Cosmology* (Garland Publishing, New York).

Hide, R. (1990). "Brief comments on George McVittie's meteorological papers," *Vistas in Astronomy 22*, 63–64.

Hill, J. A., ed. (1932). *Letters from Sir Oliver Lodge* (Cassell, London).

Holton, G. and Elkana, Y., eds. (1982). *Albert Einstein. Historical and Cultural Perspectives* (Princeton University Press, Princeton).

Howard, D. (1992). "Einstein and *Eindeutigkeit*: a neglected theme in the philosophical background to general relativity," in Eisenstaedt and Kox, eds. (1992: 154–243).

Hoyle, F. (1948). "A new model for the expanding universe," *Monthly Notices of the Royal Astronomical Society 108*, 372–382.

Hoyle, F. (1980). *Steady-State Cosmology Re-visited* (University College Cardiff Press, Cardiff).

Hoyle, F. (1982). "Steady State Cosmology revisited," in *Cosmology and Astrophysics. Essays in Honor of Thomas Gold*, Terzian, Yervant and Bilson, Elizabeth M., eds. (Cornell University Press, Ithaca), pp. 17–57.

Hoyle, F. (1994). *Home in Where the Wind Blows. Chapters from a Cosmologist's Life* (University Science Books, Mill Valley, California).

Hoyle, F. and Narlikar, J. V. (1974). *Action at a Distance in Physics and Cosmology* (W. H. Freeman, San Francisco).

Hunt, B. J. (1988). 'The origins of the FitzGerald contraction,' *British Journal for the History of Science 21*, 67–76.

Hunt, B. J. (1986). "Experimenting on the ether: Oliver J. Lodge and the Great Whirling Machine." *Historical Studies in the Physical and Biological Sciences 16*, 111–131.

Hunt, B. J. (1994). *The Maxwellians* (Cornell University Press, Cornell).

Illy, Joseph (1981). "Revolutions in a revolution," *Studies in History and Philosophy of Science 12*, 173–210.

Jeans, J. H. (1907). *The Mathematical Theory of Electricity and Magnetism* (Cambridge University Press, Cambridge).

Jeans, J. H. (1914). *The Mathematical Theory of Electricity and Magnetism* (Cambridge University Press, Cambridge).

272 *Einstein's Relativity in Great Britain*

Jeans, J. H. (1920). *The Mathematical Theory of Electricity and Magnetism* (Cambridge University Press, Cambridge).

Jeans, J. H. (1925–1926). "President's address," *Royal Astronomical Society. Monthly Notices 86*, 262–269.

Jeffreys, H. (1923). "*Space, Time, Matter*. By Hermann Weyl," *Mind 30*, 103–105.

Johnston, W. J. and Larmor, J. (1920). "The limitations of Relativity," in *Report of the Eighty-Seventh Meeting of the British Association for the Advancement of Science* (John Murray, London), pp. 11–159.

Kennefick, D. (2019). *No Shadow of a Doubt. The 1919 Eclipse that Confirmed Einstein's Theory of Relativity* (Princeton University Press, Princeton).

Kerszberg, P. (1989). *The Invented Universe. The Einstein-De Sitter Controversy (1916–1917) and the Rise of Relativistic Cosmology* (Clarendon Press, Oxford).

Kilmister, C. W. (1966). *Sir Arthur Eddington* (Pergamon Press, Oxford).

Knighting, E. (1990). "War work," *Vistas in Astronomy 33*, 59–62.

Kox, A. J., ed. (2008). *The Scientific Correspondence of H. A. Lorentz*, vol. I (Springer, New York).

Kragh, H. (1987). "The beginning of the world: Georges Lemaître and the expanding universe." *Centaurus 32*, 114–139.

Kragh, H. (1993). "Steady state theory," in Hetherington, ed. (1993: 629–636).

Kragh, H. (1996). *Cosmology and Controversy* (Princeton University Press, Princeton).

Kragh, H. (1999). "Steady-state cosmology and general relativity: Reconciliation or conflict?," in Goenner, Renn and Sauer., eds. (1999: 377–402).

Kragh, H. (2011). *Higher Speculations. Gran Theories and Failed Revolutions in Physics and Cosmology* (Oxford University Press, Oxford,).

Kransinski, A. (1990). "Early inhomogeneous cosmological models in Einstein's theory," in Bertotti, Balbinot, Bergia and Messina, eds. (1990: 115–127).

Krauss, C. R. and Laino, L., eds. (2023). *Philosophers and Einstein's Relativity. The Early Philosophical Reception of the Relativistic Revolution* (Springer, Cham).

Kuhn, T. S., Heilbron, J. L., Forman, P. and Allen, L. (1967). *Sources for History of Quantum Physics. An Inventory and Report* (The American Philosophical Society, Philadelphia).

Lakatos, I. (1970). "Falsification and the Methodology of Scientific Research Programmes," in *Criticism and the Growth of Knowledge*. I Lakatos and A. Musgrave. eds. (Cambridge University Press, Cambridge), pp. 91–195.

Reprinted: *Philosophical Papers*. J. Worrall and G. Currie, eds., vol. 1 Cambridge: Cambridge University Press (1978: 8–101).

Larmor, J. (1884). "On least action as the fundamental formulation in dynamics and physics," *Proceedings of the London Mathematical Society 15*, 158–184. Reprinted in Larmor (1929a: 31–70).

Larmor, J. (1900a). *Aether and Matter* (Cambridge University Press, Cambridge).

Larmor, J. (1900b). "The methods of mathematical physics," in *Report of the 70th Meeting of the British Association for the Advancement of Science held at Bradford in September 1900* (John Murray, London), pp. 613–629. Reprinted in Larmor (1929b: 192–216).

Larmor, J. (1918). "On the essence of Physical Relativity," *Proceedings of the National Academy of Sciences 4*, 334–337.

Larmor, J. (1919a). "On generalized Relativity in connexion with Mr. W. J. 'Johnston's Calculus," *Proceedings of the Royal Society 96*, 334–363.

Larmor, J. (1920). "Gravitation and Light." *Proceedings Cambridge Philosophical Society 19*, 324–344.

Larmor, J. (1923a). "On the nature and amount of the gravitational deflection of light," *Philosophical Magazine 45*, 243–256.

Larmor, J. (1923b). "Can gravitation really be absorbed into the frame of space and time?" *Nature* (10 February), 200.

Larmor, J. (1929a). *Mathematical and Physical Papers*, vol. I (Cambridge University Press, Cambridge).

Larmor, J. (1929b). *Mathematical and Physical Papers*, vol. II (Cambridge University Press, Cambridge).

Lecat, M. (1924). *Bibliographie de la relativité* (Maurice Lamertin, Bruxelles).

Lemaître, G. (1927). "Un univers homogene de masse constante et de rayon croissant, rendant compte de la vitesse radiale des nébuleuses extragalactiques," *Société Scientifique de Bruxelles. Annales A* 47: 49–59.

Lemaître, G. (1931). "Homogeneous universe of constant mass and increasing radius accounting for the radial velocity of extra-galactic nebulae," *Royal Astronomical Society. Monthly Notices 91*, 483–490. English translation of Lemaître (1927).

Le Roux, J. (1922). "La courbure de l'espace," *Comptes Rendus de l'Académie des Sciences 174*, 924–927.

Le Roux, J. (1922b). "Sur la gravitation dans la mécanique classique et dans la théorie d'Einstein," *Comptes Rendus de l'Académie des Sciences 175*, 809–811.

274 *Einstein's Relativity in Great Britain*

Le Roux, J. (1922c). "La mécanique de Newton n'est pas une approximation de celle d'Einstein," *Comptes Rendus de l'Académie des Sciences 175*, 1395–1397.

Levi-Civita, T. (1929a). "Vereinfachte Herstellung der Einstenschen einheitlichen Feldgleichungen," *Preussische Akademie der Wissenschaften. Phys.-mathe. Klasse, Sitzungsberichte*, pp. 137–153.

Levi-Civita, T. (1929b). "A proposed modification of Einstein's field theory," *Nature 123*, 678–679.

Lilley, A. E. (1957). "Radio astronomical measurements of interest to cosmology," in DeWitt, ed. (1957: 55–59).

Lindeman, A. F. and Lindeman, F. A. (1916–1917). "Daylight photography of stars as a means of testing the equivalence postulate in the theory of relativity," *Monthly Notices of the Royal Astronomical Society 77*, 140–151.

Lodge, O. (1892). "The motion of the ether near the Earth," *Proceedings of the Royal Institution 13*, 565–580.

Lodge, O. (1893). "Aberration problems. A discussion concerning the motion of the ether near the Earth, and concerning the connexion between ether an gross matter; with some new experiments," *Philosophical Transactions of the Royal Society* A *184*, 727–804.

Lodge, O. (1917a). "Astronomical consequences of the electrical theory of matter," *Philosophical Magazine 34*, 81–94.

Lodge, O. (1917b). "Astronomical consequences of the electrical theory of matter. Supplementary note," *Philosophical Magazine 34*, 517–521.

Lodge, O. (1918). "Continued discussion of the astronomical and gravitational bearings of the electrical theory of matter," *Philosophical Magazine 35*, 141–156.

Lodge, O. (1919). "The new theory of gravity," *The Nineteenth Century and After 86*, 1189–1200.

Lodge, O. (1920). "Note on a possible structure for the ether," *Philosophical Magazine 39*, 170–174.

Lodge, O. (1931). *Advancing Science Being Personal Reminiscences of the British Association in the Nineteenth Century* (Ernest Benn, London).

Lodge, O. (1932). *Past Years. An Autobiography* (Charles Scribner's, New York).

Lodge, O. (1933). *My Philosophy, Representing My Views on the Many Functions of the Ether of Space* (Ernest Benn, London).

Lorentz, H. A. (1875). *Over de theorie der terugkaatsing en breeking van het licht* (Academisch proefschrift, Leiden). Reproduced in Lorentz (1935: 1–192).

Lorentz, H. A. (1892). "The relative motion of the Earth and the ether," *Versl. K. Akad. W. Amsterdam 1*, 74–79. Reproduced in Lorentz (1937a: 219–223).

Lorentz, H. A. (1895). *Versuch einer Theorie der electrischen und optischen Erscheinungen in bewegten Körpern* (Brill, Leiden). Reproduced in Lorentz (1937b: 1–137).

Lorentz, H. A. (1897). "Concerning the problem of the dragging along of the ether by the Earth," *Versl. K. Akad. W. Amsterdam 6*, 226–274. Reproduced in Lorentz (1937a: 237–244).

Lorentz, H. A. (1900). *The Theory of Electrons and its Applications to the Phenomena of Light and Radiant Heat* (Teubner, Leipzig).

Lorentz, H. A. (1904). Electromagnetic phenomena in a system moving with any velocity smaller than that of light," *Proceedings Royal Academy of Amsterdam 6*, 809–. Reproduced in Lorentz (1937b: 172–197).

Lorentz, H. A. (1935). *Collected Papers*, vol. I (Martinus Nijhoff, The Hague).

Lorentz, H. A. (1937a). *Collected Papers*, vol. IV (Martinus Nijhoff, The Hague).

Lorentz, H. A. (1937b). *Collected Papers*, vol. V (Martinus Nijhoff, The Hague).

Lorentz, H. A. (1952). *The Theory of Electrons and its Applications to the Phenomena of Light and Radiant Heat* (Dover, New York).

Lowe, V. (1985). *Alfred North Whitehead: The Man and His Work*, vol. 1 (*1861–1910*) (The Johns Hopkings University Press, Baltimore).

Lowe, V. (1990). *Alfred North Whitehead: The Man and His Work*, vol. 2 (*1910–1941*) (The Johns Hopkings University Press, Baltimore).

Maiocchi, R. (1985). *Einstein in Italia. La scienza e la filosofia italiane di fronte alla teoria della relatività* (Franco Angeli, Milano).

Maurice, F. (1937). *Haldane: The Life of Viscount Haldane of Cloan, K.T., O.M.* (Faber and Faber Limited, London).

Mavridès, S. (1973). *L'Univers relativiste* (Masson, Paris).

McCrea, W. H. (1923). "Cosmology. A brief review." *Quarterly Journal of the Royal Astronomical Society 4*, 185–201.

McCrea, W. H. (1990). "George Cunliffe McVittie (1904–1988), OBE, FRSE, pupil of Whittaker and Eddington: pioneer of modern cosmology," *Vistas in Astronomy 33*, 43–58.

McLaren, S. B. (1913). "A theory of gravity," *Philosophical Magazine 26*, 636–673.

McVittie, G. C. (1929a). "On Einstein's unified field theory," *Proceedings of the Royal Society of London A 124*, 366–374.

McVittie, G. C. (1929b). "On Levi-Civita's modification of Einstein's unified field theory," *Philosophical Magazine 8*, 1033–1044.

McVittie, G. C. (1932). "Condensations on an expanding universe," *Monthly Notices of the Royal Astronomical Society 92*, 500–518.

276 *Einstein's Relativity in Great Britain*

McVittie, G. C. (1933). "The mass-particle in an expanding universe," *Monthly Notices of the Royal Astronomical Society 93*, 325–339.

McVittie, G. C. (1935a). "Absolute parallelism and Milne's kinematical relativity," *Monthly Notices of the Royal Astronomical Society 95*, 270–279.

McVittie, G. C. (1935b). "Absolute parallelism and metric in the expanding universe theory," *Proceedings of the Royal Society of London A 151*, 357–370.

McVittie, G. C. (1937). *Cosmological Theory* (Methuen, London 1937).

McVittie, G. C. (1940). "Kinematical relativity," *The Observatory 63*, 273.

McVittie, G. C. (1941). "Kinematical relativity — a discussion (with E. A. Milne and A. G. Walker)," *The Observatory 64*, 11.

McVittie, G. C. (1956). *General Relativity and Cosmology* (Chapman and Hall, London).

McVittie, G. C. (1957). "Counts of extra-galactic radio sources and uniform model universes," *Australian Journal of Physics 10*, 331–350.

McVittie, G. C. (1958). "Distance and relativity," *Science 127*, 501–505.

McVittie, G. C. (1959). "Distance and time in cosmology: the observational data," *Handbuch der Physik 53*, 445 (Springer-Verlag, Berlin).

McVittie, G. C. (1961). *Fact and Theory in Cosmology* (Eyre & Spottiswoode, London).

McVittie, G. C. (1962a). "Galaxies as members of the Universe: Summary," in McVittie, ed. (1962), pp. 441–450

McVittie, G. C. (1962b). "Cosmology and the interpretation of astronomical data," in Royaumont (1962), pp. 253–274.

McVittie, G. C. (1965a). *General Relativity and Cosmology* (Chapman and Hall, London), second edition.

McVittie, G. C. (1965b). "Some consequences of large redshifts," *Astrophysical Journal 142*, 1637.

McVittie, G. C. (1974). "Distance and large redshifts," *Quarterly Journal of the Royal Astronomical Society 15*, 246.

McVittie, G. C. (ca. 1975). "Autobiographical sketch," prepared at the request of the Royal Society of Edinburgh, copy deposited at the Niels Bohr Library, American Institute of Physics.

McVittie, G. C. (1978). *Transcript of an Interview [made by David DeVorkin] taken on a Tape Recorder* (American Institute of Physics, Center for History of Physics, New York).

McVittie, G. C. (1987). "An Anglo-Scottish university education," in Williamson, ed. (1987), pp. 66–70.

References 277

McVittie, G. C., ed. (1962). *Problems of Extra-Galactic Research* (The Macmillan Co., New York).

McVittie, G. C. and McCrea, W. H. (1931). "On the contraction of the universe," *Monthly Notices of the Royal Astronomical Society 91*, 128–133.

Mercier, A. (1992). "General relativity at the turning point of its renewal," in Eisenstaedt and Kox, eds. (1992: 109–121).

Mercier, A. and Kervaire, M., eds. (1956). *Fünfzig Jahre Relativitätstheorie, Helvetica Physica Acta*, Supplement IV (Birkhäuser Verlag, Basel).

Merz, J. T. (1904). *A History of European Thought in the Nineteenth Century*, vol. 2. (William Blackwood, London). Reprinted under the title *A History of European Scientific Thought in the Nineteenth Century*, vol. 2 (Dover, New York 1965).

Metz, R. (1938). *A Hundred Years of British Philosophy*, 2 vols. (The Macmillan Co., New York). English translation of *Die philosophischen Strömungen der Gegenwart in Grossbritannien* (1935).

Meyerson, É. (1925). *La deduction relativiste* (Payot, Paris).

Michelson, A. A. and Morley, E. W. (1887). "On the relative motion of the Earth and the luminiferous ether," *The American Journal of Physics 34*, 333–345.

Milne, E. A. (1935). *Relativity, Gravitation and World-Structure* (Clarendon Press, Oxford).

Milne, E. A. (1940). "Kinematical relativity," *Journal of the London Mathematical Society 15*, 44–80.

Milne, E. A. (1943). *Kinematical Relativity; a Sequel to Relativity, Gravitation and World-Structure* (Clarendon Press, Oxford).

Milne, E. A. (1952). *Modern Cosmology and the Christian Idea of God* (Clarendon Press Oxford).

Minkowski, H. (1909). "Raum und Zeit," *Physikalische Zeitschrift 20*, 104–111.

Missner, M. (1985). "Why Einstein Became Famous in America," *Social Studies of Science 15*, 267–291.

Møller, C. (1962). "The energy-momentum complex in general relativity and related problems," in Royaumont (1962), pp. 15–29.

North, J. (1990). *The Measure of the Universe* (Dover, New York).

North, J. (1994). *The Fontana History of Astronomy and Cosmology* (Fontana Press, London).

Nunn, P. (1923). *Relativity & Gravitation* (University of London Press, London).

O'Raifeartaigh, L., ed. (1972). *General Relativity. Papers in Honour of J. L. Synge.* (Clarendon Press, Oxford).

278 *Einstein's Relativity in Great Britain*

Passmore, J. (1966). *A Hundred Years of Philosophy* (Duckworth, London; first edition of 1957).

Palter, R. M. (1960). *Whitehead's Philosophy of Science* (Chicago, Illinois).

Paty, M. (1987). "The scientific reception of relativity in France," in Glick, ed. (1987: 113–167).

Payne-Gaposchkin, C. (1925). *Stellar Atmospheres* (W. Heffer & Sons, Cambridge).

Penrose, R. (1965). "Gravitational collapse and space-time singularities," *Physical Review Letters 14*, 57–59.

Penrose, R. (c. 1966). *An Analysis of the Structure of Space-Time*. Adams Prize essay.

Penrose, R. (1968). "Structure of Space-Time," in *Battelle Rencontres. 1967 Lectures in Mathematics and Physics*, DeWitt, C., M. and Wheeler, J. A., eds. (W. A. Benjamin, New York), pp. 121–235.

Penrose, R. (2011). "Biography. Early years and influences," in *Roger Penrose Collected Works*, vol. 6 (*1997–2003*) (Oxford University Press, Oxford), pp. xi–xii.

Penzias, A. A., and Wilson, R. W. (1965). "A measurement of excess antenna temperature at 4080 Mc/s," *Astrophysical Journal 142*, 419–421.

Pirani, F. A. E. (1956). "On the physical significance of the Riemann tensor," *Acta Physica Polonica 15*, 389–405.

Planck, M. (1930). "Theoretische Physik," in *Aus 50 Jahren deutscher Wissenschaft. Die Entwicklung ihrer Fachgebiete in EinseldarsteUungen (Schmidt-Ott Festschrift)* (W. de Gruyter, Berlin), pp. 300–209.

Planck, M. (1932). "Fifty years of science," in *Where is Science Going?* (W. W. Norton, New York), pp. 41–63.

Poincaré, H. (1906). "Sur la dynamique de l'électron." *Rendiconti del Circolo Matematico di Palermo 21*, 129–175. Reprinted in *Oeuvres de Henri Poincaré*, vol. 9, G. Petiau, ed. (Gauthier-Villars, Paris 1954: 494–550.

Popper, K. (1974). "Autobiography," in Schilpp, ed. (1979), vol. I, pp. 1–181.

Poynting, J. H. (1920). *Collected Scientific Papers* (Cambridge University Press, Cambridge), pp. 599–612. Reprinted from *British Association Report 1899*, pp. 61–24.

Preston, T. (1890). *The Theory of Light* (Macmillan, London).

Pyenson, L. (1987). "The relativity revolution on Germany," in Glick, ed. (1987: 59–111).

Rabinovici, E. (2020). "Jacob Bekenstein (1947–2015): A conservative revolutionary," in *Jacob Bekenstein. The Conservative Revolutionary*, Brink, Lars,

Mukhanov, Viatcheskav, Rabinovici, Eliezer and Phua, K. K., eds. (World Scientific, Singapore), pp. xi–xiii.

Reeves, B. J. (1987). "Einstein politicized: the early reception of relativity in Italy," in Glick (1987: 189–229).

Reichenbach, H. (1949). "The philosophical significance of the theory of relativity," in SCHILPP, ed. (1949: 287–311).

Reichenbach, H. (1978). *Selected Writings, 1909–1953*, Maria Reichenbach and Robert S. Cohen, eds., vol. 1 (Reidel, Dordrecht).

Renn, J. S. M., eds (2007). *The Genesis of General Relativity*, vol. 3 (*Gravitation in the Twilight of Classical Physics*) (Springer, Dordrecht).

Report of the Eighty-Sixth Meeting of the British Association for the Advancement of Science: Newcastle-On-Tyne: 1916 (1917) (John Murray, London).

Review Phil. Mag. (1923). "Whitehead's *The Principle of Relativity*," *Philosophical Magazine 45*, 1103.

Rice, J. (1923). *Relativity: A Systematic Treatment of Einstein's Theory* (Longmans, Green, London).

Richardson, O. W. (1922). "Problems of Physics," in *Report of the Eighty-Ninth Meeting of the British Association for the Advancement of Science* (John Murray, London), pp. 25–35.

Robb, A. A. (1911). *Optical Geometry of Motion: a New View of the Theory of Relativity* (Heffner and Sons, Cambridge).

Robb, A. A. (1914). *A Theory of Time and Space* (Cambridge University Press, Cambridge).

Robb, A. A. (1921). *The Absolute Relations of Time and Space* (Cambridge University Press, Cambridge).

Robb, A. A. (1936). *Geometry of Time and Space* (Cambridge University Press, Cambridge).

Robertson, H. P. (1928). "On relativistic cosmology," *Philosophical Magazine 5*, 835–848.

Robertson, H. P. (1929). "On the foundations of relativistic cosmology," *Proceedings National Academy of Sciences 15*, 822–829.

Robinson, A. (2019). *Einstein on the Run. How Britain Saved the World's Greatest Scientist* (Yale University Press, New Haven).

Robinson, D. C. (2019). "Gravitation and general relativity at King's College London," *The European Physical Journal H 44*, 181–270.

Ross, W. D. (1921). "Book reviews," *Mind 30*, 232–233.

Rotenstreich, Nathan (1982). "Relativity and relativism," in Holton and Elkana, eds. (1982: 175–204).

280 *Einstein's Relativity in Great Britain*

Roxburgh, I. W. (2007). "Sir Hermann Bondi KCB, 1 November 1919–10 September 2005)," *Biographical Memoirs of Fellows of the Royal Society 53*, 45–60.

Royaumont (1962). *Les théories relativistes de la gravitation* (Éditions du Centre National de la Recherche Scientifique, Paris).

Russell, B. (1914). *Our Knowledge of the External World as a Field for Scientific Method in Philosophy* (Open Court, Chicago).

Russell, B. (1917). "Review of May Sinclair's *Defence of Idealism: Some Questions and Conclusions*," *The Nation 21* (8 September), pp. 588, 590. Reprinted in Russell (1986: 106–110).

Russell, B. (1919). "Einstein's theory of relativity," *The Atheneum*, no. 4,672 (14 November), 1,189. Reprinted in Russell (1988: 207–209).

Russell, B. (1922). "Physics and perception," *Mind 31*, 478–485.

Russell, B. (1926). "Relativity: Philosophical consequences," *Encyclopedia Britannica*, 13th edition (London), pp. 331–332. Reprinted in Russell (1988: 228–234).

Russell, B. (1968). *The Autobiography of Bertrand Russell*, vol. II ("1914–1944") (George Allen and Unwin LTD., London).

Russell, B. (1986). *The Philosophy of Logical Atomism and Other Essays, 1914–1919* (Unwin Hyman, London).

Russell, B. (1988). *Essays on Language, Mind and Matter, 1919–1926* (Unwin Hyman, London).

Russell, H. N., Dugan, R. S. and Stewart, J. Q. (1926–1927). *Astronomy*, 2 vols. (Ginn and Co., Boston).

Rutherford, E. and Compton, A. H. (1919). "Radio-activity and gravitation," *Nature 104*, 412.

Ryckman, T. (2005). *The Reign of Relativity. Philosophy in Physics, 1915–1925* (Oxford University Press, Oxford).

Ryle, M. (1955). "Radio stars and their cosmological significance," *The Observatory 75*, 137–147.

Sadler, D. H. (1987). "The decade 1940–1950," in Tayler, ed. (1987: 98–147).

Sanchez-Ron, J. M. (1987a). "The reception of special relativity in Great Britain," in Glick, ed. (1987: 27–58).

Sanchez-Ron, J. M. (1987b). "The role played by symmetries in the introduction of relativity in Great Britain," in *Symmetries in Physics (1500–1980)*, G. Doncel, Manuel, Hermann, Armin, Michel, Louis and Pais, Abraham, eds. (Seminari d'História de les Ciènces, Universitat Autónoma de Barcelona, Bellaterra)), pp. 165–184.

Sanchez-Ron, J. M. (1990). "Steady-state theory, the arrow of time, and Hoyle and Narlikar's theories," in Bertotti, Balbinot, Bergia and Messina, eds. (1990: 233–243).

Sanchez-Ron, J. M. (1992). "The reception of general relativity among British physicists and mathematicians (1915–1930)," *Einstein Studies*, vol. 3 (Birkhäuser, Boston), pp. 57–88.

Sanchez-Ron, J. M. (1999). "Larmor versus general relativity," in *The Expanding Worlds of General Relativity*, vol. 7 of *Einstein Studies* (Birkhäuser, Boston), pp. 405–430.

Sanchez-Ron, J. M. (2005). "George McVittie, the uncompromising empiricist," in *The Universe of General Relativity*, vol. 11 of *Einstein Studies* (Birkhäuser, Boston), pp. 189–221.

Sanchez-Ron, J. M. (2012). "The early reception of Einstein's relativity among British philosophers," in *Einstein and the Changing World Views of Physics*, Lehner, C., Renn, J. and Schemmel, M., eds. (Birkhäuser, Boston), pp. 75–118.

Schaffner, K. F. (19712). *Nineteenth-Century Aether Theories* (Pergamon Press, Oxford).

Schilck, M. (1915). "Die philosophische Bedeutung des Relativitätsprinzip," *Zeitschrift für Philosophie und philosophische Kritik 159*, 129–175. English translation Schilck (1979b).

Schilck, M. (1917a). "Raum und Zeit in der gegenwärtigen," *Die Naturwissenschften* 5, 161–167, 177–186.

Schilck, M. (1917b). *Raum und Zeit in der gegenwärtigen* (Julius Springer Verlag, Berlin).

Schilck, M. (1918). *Allgemeine Erkenntnislehre* (Julius Springer, Berlin).

Schilck, M. (1922). "Die Relativitätstheorie in der Philosophie," *Verhandlungen der Gesellschaft Deutscher Naturforscher und Ärzte. 87. Versammlung, Hundertjahrfeier* (Leipszig), pp. 58–69. English translation Schlick (1979c).

Schilck, M. (1974). *General Theory of Knowledge* (Springer-Verlag, Wien); English translation of the second edition (1925).

Schilck, M. (1979a). *Moritz Schlick: Philosophical Papers*, H. L. Mulder and B. van de Velde-Schlick, eds., vol. 1 (Reidel, Dordrecht).

Schilck, M. (1979b). "The philosophical significance of the principle of relativity, in Schlick (1979a: 153–189).

Schilck, M. (1979c). "The theory of relativity in philosophy," in Sclick (1979a: 343–353).

282 *Einstein's Relativity in Great Britain*

Schilpp, P. A. ed. (1949). *Albert Einstein: Philosopher-Scientist* (Open Court, La Salle, Ill.).

Schilpp, P. A. ed. (1959). *The Philosophy of C. D. Broad* Open Court, La Salle, Ill.).

Schilpp, P. A. ed. (1963). *The Philosophy of Rudolf Carnap* (Open Court, La Salle, Ill.).

Schilpp, P. A. ed. (1974). *The Philosophy of Karl Popper* (Open Court, La Salle, Ill.).

Schott, G. A. (1907a). "On the Radiation from Moving Systems of Electrons, and on the Spectrum of Canal Rays," *Philosophical Magazine 13*, 657–687.

Schott, G. A. (1907b). 'On the Electron Theory of Matter and of Radiation,' *Philosophical Magazine 13*, 189–213.

Schott, G. A. (1912). *Electromagnetic Radiation* (Cambridge University Press, Cambridge).

Schuster, A. (1911). *The Progress of Physics during 33 years (1875–1908)* (Cambridge University Press, Cambridge).

Schweber, S. S. (1994). *QED and the Men Who Made It: Dyson, Feynman, Schwinger, and Tomonaga* (Princeton University Press, Princeton).

Schild, A. (1956). "On gravitational theories of Whitehead's type," *Proceedings of the Royal Society of London A 235*, 202–209.

Schild, A. (1962a). "Gravitational theories of the Whitehead type and the principle of equivalence," in *Rendiconti della Scuola Internazionale di Fisica "Enrico Fermi," XX Corso, "Verifiche delle teorie gravitazionali"* (Academic Press, New York), pp. 69–115.

Schild, A. (1962b). "The principle of equivalence," *The Monist 47*, No. 1, 20–39.

Schild, A. (1962c). "Conservative gravitational theories of Whitehead's type," in *Recent Developments in General Relativity* (Pergamon Press-Polish Scientific Publishers, Oxford-Warsaw), pp. 409–413.

Sciama, D. (1953). "On the origin of inertia," *Monthly Notices of the Royal Astronomical Society 113*, 34–42.

Sciama, D. (1959). *The Unity of the Universe* (Faber and Faber, London).

Sciama, D. (1969). *The Physical Principles of General Relativity* (Doubleday, New York).

Sciama, D. (1976). "Black holes and their thermodynamics," *Vistas in Astronomy 19*, 385–401.

Senechal, M. (2013). *I Died for Beauty. Dorothy Wrinch and the Cultures of Science* (Oxford University Press, Oxford).

Silberstein, L. (1914). *The Theory of Relativity* (Macmillan and Co., London).

Silberstein, L. (1930). *The Size of the Universe. Attempts at a Determination of the Curvature Radius of Spacetime* (Oxford University Press, Oxford).

Silberstein, L. (1918a). "Bizarre conclusion derived from Einstein's Gravitation Theory," *Royal Astronomical Society. Monthly Notices 78*, 465–467.

Silberstein, L. (1918b). "General Relativity without the Equivalence Hypothesis," *Philosophical Magazine 36*, 94–128.

Silberstein, L. (1920). "The recent eclipse results and Stokes-Planck's aether," *Philosophical Magazine 39*, 161–170.

Silberstein, L. (1922). *The Theory of General Relativity and Gravitation* (Van Nostrand, New York).

Silberstein, L. (1930). *The Size of the Universe. Attempts at a Determination of the Curvature Radius of Spacetime* (Oxford University Press).

Slater, J. G. (1988). "Introduction," in Russell (1988: xiii–xxv).

Spirito, U. (1921). "Le interpetrazioni [*sic*] idealistiche delle teorie di Einstein," *Giornale critico della filosofia italiana 2*, no. 2, 63–75.

Stachel, J. (1986). "Eddington and Einstein," in *The Prism of Science*, E. Ullmann-Margalit, ed. (Reidel, Dordrecht), pp. 225–250. Reprinted in Stachel (2002), pp. 453–475.

Stachel, J. (2002). *Einstein from 'B' to 'Z'* (Birkhäuser, Boston).

St. John, C. E. (1922–1923). "On gravitational displacement of solar lines," *Royal Astronomical Society. Monthly Notices 84*, 93–96.

Stanley, M. (2007). *Practical Mystic. Religion, Science and A. S. Eddington* (The University of Chicago Press, Chicago).

Stanley, M. (2020). "Lodge and mathematics. Counting beans, the meaning of symbols, and Einstein's blindfold," in *A Pioneer of Connection. Recovering the Life and Work of Oliver Lodge*, Mussell, James and Gooday, eds. (University of Pittsburgh Press, Pittsburgh), pp. 87–103.

Strong, C. A. (1922). "The meaning of meaning," *Mind 31*, 69–71.

Synge, J. L. (1951). *The relativity theory of A. N. Whitehead*. Lecture Series No. 5, Institute for Fluid Dynamics and Applied Mathematics, University of Maryland.

Synge, J. L. (1952). "Orbits and rays in the gravitational field of a finite sphere according to the theory of A. N. Whitehead," *Proceedings of the Royal Society of London A 211*, 303–319.

Synge, J. L. (1958). "Whittaker's contributions to the theory of Relativity," *Proceedings of the Edinburgh Mathematical Society 11*, 39–55.

Synge, J. L. (1960). *Relativity: The General Theory* (North-Holland, Amsterdam).

284 *Einstein's Relativity in Great Britain*

Tanaka, Y. (1987). "Einstein and Whitehead. The Principle of Relativity reconsidered," *Historia Scientiarum*, No. 32, 43–61.

Tayler, R. J., ed. (1987). *History of the Royal Astronomical Society*, vol. 2 (*1920–1980*) (Blackwell, Oxford).

Taylor, A. E. (1921). "*Relativity, the Special and General Theory: A Popular Exposition*. By Albert Einstein; *Space, Time and Gravitation: An Outline of the General Theory of Relativity*. By A. S. Eddington; *The Concept of Nature*: Tarner Lectures delivered in Trinity College, November, 1919. By A. N. Whitehead," *Mind 30*, 76–83.

Temple, G. (1926). "A theory of relativity in which the dynamical manifold can be conformally represented upon the metrical manifold," *Proceedings of the London Mathematical Society 25*, 401–416.

Temple, G. (1956). "Edmund Taylor Whittaker, 1873–1956," *Biographical Memoirs of Fellows of the Royal Society 2*, 299–325.

Thomson, J. J. (1910). "President's address," in *Report of the Seventy-Ninth Meeting of the British Association for the Advancement of Science, Winnipeg, 2900, August 25-September 1* (John Murray, London), pp. 3–29.

Thomson, J. J. (1936). *Recollections and Reflections* (G. Bell and Sons, London).

Turner, J. E. (1916). "The nature and geometry of space," *Mind 25*, 223–228.

Turner, J. E. (1920). "Relativity, scientific and philosophical," *Mind 31*, 337–342.

Turner, J. E. (1922). "Dr. Wildom Carr and Lord Haldane on scientific relativity," *Mind 31*, 40–42.

Urani, J. and Gale, G. (1993). "E. A. Milne and the origins of modern cosmology: an essential presence," in Earman, Janssen and Norton, eds. (1993: 390–419).

Vucinich, A. (2001). *Einstein and Soviet Ideology* (Stanford University Press, Stanford).

Walker, G. W. (1918). "Relativity and electrodynamics," *Philosophical Magazine 35*, 327–338.

Warwick, A. (1993). "Frequency, theorem and formula: remembering Joseph Larmor in electromagnetic theory," *Notes and Records of the Royal Society of London 47*, 49–60.

Warwick, A. (2003). *Masters of Theory. Cambridge and the Rise of Mathematical Physics* (The University of Chicago Press Chicago).

Weaire, D. (2009). "Watch this space: the physics of an empty box," en Weaire, D., ed., *George Francis Fitzgerald* (Living Edition, Pollauberg 2009), pp. 129–136.

References 285

Whittaker, E. T. (1927). "The outstanding problems of Relativity." *Report of the Ninety-Fifth Meeting of the British Association for the Advancement of Science* (Office of the British Association, London), pp. 16–26.

Whittaker, E. T. (1929). "What is Energy?," *Mathematical Gazette* (April): 401–406.

Whittaker, E. T. (1953). *A History of the Theories of Aether and Electricity. II. The Modem Theories, 1900–1926* (Thomas Nelson and Sons, London and Edinburgh).

Whitehead, A. N. (1915–1916). "Space, time and Relativity," *Proceedings of the Aristotelian Society*, 104–129.

Whitehead, A. N. (1919). *An Enquiry Concerning the Principles of Natural Knowledge* (Cambridge University Press, Cambridge).

Whitehead, A. N. (1920). *The Concept of Nature* (Cambridge University Press, Cambridge).

Whitehead, A. N. (1922). *The Principle of Relativity with Applications to Physical Science* (Cambridge University Press, Cambridge).

Williamson, R., ed. (1987). *The Making of Physicists* (Adam Hilger, Bristol).

Wilson, M. (1951). *Ninth Astronomer Royal. The Life of Frank Watson Dyson* (W. Heffer & Sons, Cambridge).

Wilson, W. (1919). "Relativity and gravitation," *Physical Society of London. Proceedings 31*, 69–78.

Woodruff, A. E. (1981). "Larmor, Joseph," in *Dictionary of Scientific Biography*. Vol. 8, C. Gillespie, ed. (Scribner & Sons, New York), pp. 39–41.

Wright, H., Warnow, J. N. and Weiner, C., eds. (1972). *The Legacy of George Ellery Hale* (The MIT Press, Cambridge, Mass.).

Wrinch, D. (1922). "On certain methodological aspects of the theory of relativity," *Mind 31*, 200–204.

Wrinch, D. (1924). "*Scientific Thought. By C. D. Broad*," *Mind 33*, 184–192.

Wrinch, D. (1927). "The relations of science and philosophy," *Journal of Philosophical Studies 2*, 153–166.

Wrinch, D. (1929–1930). "Scientific method in some embryonic sciences," *Proceedings of the Aristotelian Society*, 229–242.

Wrinch, D. (1937). "The cyclol hypothesis and the 'globular proteins'," *Proceedings of the Royal Society of London A 161*, 505–524.

Wrinch, D. and Jeffreys, H. (1921). "The relation between geometry and Einstein's theory of gravitation," *Nature 106*, 806–809.

www.ingramcontent.com/pod-product-compliance
Lightning Source LLC
Chambersburg PA
CBHW050352090625

27790CB00004B/19